Lecture Notes in Mathematics 1571

Sergei Yu. Pilyugin

The Space of Dynamical Systems with the C^0-Topology

Springer-Verlag
Berlin Heidelberg New York
London Paris Tokyo
Hong Kong Barcelona
Budapest

Author

Sergei Yu. Pilyugin
Department of Mathematics and Mechanics
St. Petersburg State University
Bibliotechnaya pl.
2, Petrodvorets
198904 St. Petersburg, Russia

Mathematics Subject Classification (1991): 58F30, 58F10, 58F12, 58F40, 54H20, 65L99

ISBN 3-540-57702-5 Springer-Verlag Berlin Heidelberg New York
ISBN 0-387-57702-5 Springer-Verlag New York Berlin Heidelberg

Printed in Germany

SPIN: 10078801 46/3140-543210 - Printed on acid-free paper

Preface

A standard object of interest in the global qualitative theory of dynamical systems is the space of smooth dynamical systems with the C^1-topology. In recent years many deep and important results were obtained in the theory of structural stability. These results are mostly based on the following fundamental fact : we may consider a C^1-small perturbation of a smooth dynamical system as a perturbation in a neighborhood of any trajectory which does not change essentially the corresponding "first approximation" linear system. It is known for a long time (beginning with works of A.Lyapunov and H.Poincaré) that under some intrinsic conditions on the "first approximation" system perturbations of this sort do not change the local structure of a neighborhood of a trajectory.

The situation becomes quite different if we study C^0-small perturbations of a system. It is easy to understand that arbitrarily C^0-small perturbation can result in a complete change of the qualitative behaviour of trajectories in a neighborhood of a fixed trajectory. Nevertheless the theory of C^0-small perturbations of dynamical systems which was developed intensively over the last 20 years contains now many interesting results.

It was an intention of the author to give the reader an initial perspective of the theory. So we are going to give in this book an introduction to some of the main methods of the theory and to formulate its principal results.

Of course, this book is a reflection of scientific interests of the author, hence we pay more attention to problems which are close to the author's own works. This book is an introduction rather than a monograph. That's why the author tried to simplify and to "visualize" some proofs. Due to this reason some technically complicated proofs (mostly connected with applications of the theory of systems with hyperbolic structure) are omitted, and the reader is referred to the original papers. Sometimes we give only an "explanation" of the main ideas instead of a complete proof (as for example in the case of the C^0-density Theorem of M.Shub).

The book consists of 5 chapters and 3 Appendices. Chapter 0 contains practically no theorems. It gives an introduction to the language of the theory and surveys some results we need later. In section 0.1 we define spaces of dynamical systems : the space $Z(M)$ of continuous discrete dynamical systems on a smooth closed manifold M – the main object in this book, and some other spaces we work with. We devote section 0.2 to the space M^* - the space of compact subsets of M with different topologies. We describe also some properties of semicontinuous set-valued maps in this section. In section 0.3 we prove two variants of the C^0-closing Lemma. In section 0.4 we give a survey of basic results of the theory of smooth dynamical systems with hyperbolic structure. We describe the set of diffeomorphisms which satisfy the STC (the strong transversality condition). It is known now that this set coincides with the set of structurally stable systems. We try to explain in this book that diffeomorphisms which satisfy the STC play the crucial role not only in the theory of structural stability but also in the theory of C^0-small perturbations of dynamical systems.

Chapter 1 is devoted to generic properties of systems in $Z(M)$. We study tolerance stability in section 1.1. A counterexample constructed by W.White to the Tolerance Stability Conjecture is described. We prove some results of F.Takens connected with this conjecture. Pseudoorbits are considered in section 1.2. We define the POTP (the pseudo orbit tracing property) and some of its generalizations. The genericity of weak shadowing for systems in $Z(M)$ is established. We give also a proof by F.Takens of a variant of the Tolerance Stability Conjecture (with extended orbits instead of orbits).

Various types of prolongations are studied in section 1.3. We describe results of V.Dobrynsky and A.Sharkovsky which show that a generic system in $Z(M)$ has the following property : the set of points such that their positive trajectories are stable with respect to permanent perturbations is residual in M. It is shown also that a generic system in $Z(M)$ has the property : for any point of M its prolongation with respect to the initial point, its prolongation with respect to the system, and its chain prolongation coincide.

We study various sets of returning points : the nonwandering set, the set of weakly nonwandering points, the chain-recurrent set in section 1.4. In this section we discuss also filtrations. We prove a theorem of M.Shub and S.Smale : if a system has a fine sequence of filtrations then it has no C^0 Ω-explosions.

Chapter 2 is devoted to topological stability. We describe general properties of topologically stable systems in section 2.1. It is shown that if a dynamical system ϕ is topologically stable then ϕ is tolerance-$Z(M)$-stable and ϕ has the POTP. We prove also the following result obtained by P.Walters and A.Morimoto : if a system ϕ is expansive and has the POTP then ϕ is topologically stable.

Results of Z.Nitecki on topological stability of diffeomorphisms with hyperbolic structure are described in section 2.2. We show that a hyperbolic set is locally topologically stable. After that we apply Smale's techniques for constructing filtrations to prove that if a diffeomorphism ϕ satisfies the Axiom A and the no-cycle condition then ϕ is topologically Ω-stable. We formulate (without a proof) the main result of Z.Nitecki : if a diffeomorphism ϕ satisfies the STC then ϕ is topologically stable.

K.Yano characterized topologically stable dynamical systems on the circle. The main statement of section 2.3 is the following theorem of K.Yano : a system ϕ on S^1 is topologically stable if and only if ϕ is topologically conjugate to a Morse-Smale diffeomorphism. We describe in this section an example constructed by K.Yano of a dynamical system which has the POTP but is not topologically stable.

Section 2.4 is devoted to the C^0-density theorem of M.Shub : any diffeomorphism ϕ can be isotoped to a diffeomorphism satisfying the STC by an isotopy which is arbitrarily small in the C^0-topology. We do not give a complete proof of this result but describe its main ideas in the most "visible" case $\dim M = 2$. In section 2.5 we formulate (without proofs) two results : a theorem of M.Hurley who described the chain-recurrent set of a topologically stable diffeomorphism and a theorem of J.Lewowich who applied Lyapunov type functions in the theory of topological stability.

We study C^0-small perturbations of attractors in Chapter 3. Basic properties of attractors are described in section 3.1. Section 3.2 is devoted to stability of attractors under C^0-small perturbations of the system with respect to different metrics on M^*. It is shown that a generic system ϕ in $Z(M)$ has the property : any attractor of ϕ is stable with respect to R_0. This result is a generalization of a theorem of M.Hurley which considers stability of attractors with respect to the Hausdorff metric R. We prove also the following result of the author : if an attractor is stable with respect to R_0 then its boundary is Lyapunov stable.

Lyapunov stable sets and quasi-attractors in generic systems are considered in section 3.3. We prove a theorem of M.Hurley : if $\dim M \leq 3$ then the union of basins of chain-transitive quasi-attractors of a generic dynamical system is a residual subset of M. The second main result of this section was obtained by the author : for a generic dynamical system any Lyapunov stable set is a quasi-attractor.

In section 3.4 we study stability of attractors with the STC on the boundary. The main theorem of this section was proved by O.Ivanov and the author : if I is an attractor of a diffeomorphism ϕ which satisfies the STC on the boundary of I then I is Lipschitz stable with respect to the Hausdorff metric R. Section 3.5 is devoted to stability of attractors for Morse-Smale diffeomorphisms. We describe results of V.Pogonysheva. It is shown that an attractor I of a Morse-Smale diffeomorphism is stable with respect to metric R_2 on M^* if and only if $I = \overline{\text{Int} I}$.

In Chapter 4 we study limit sets of domains and describe some results obtained by the author and V.Pogonysheva. Section 4.1 is devoted to Lyapunov stability of limit sets. It is shown, for example, that a generic system in $Z(M)$ has the following property : given $x \in M$ there exists a countable set $B(x)$ such that for any $r \in (0, +\infty) \setminus B(x)$ the ω-limit set of the ball of radius r centered at x is Lyapunov stable. We investigate also the process of "iterating of taking limit sets of neighborhoods". It is shown that for a generic system this process "approximately stops" after the first step.

Section 4.2 is devoted to limit sets for diffeomorphisms which satisfy the STC. It is shown that if a diffeomorphism ϕ satisfies the STC then given $x \in M$ there is a finite set $C(x)$ such that for any $r \in (0, +\infty) \setminus C(x)$ the ω-limit set of the ball of radius r centered at x is an attractor. We prove a result of V.Pogonysheva which gives sufficient conditions for the stability of the ω-limit set of a domain G with respect to both the set G and to the system.

Appendices A,B of the book are devoted to two important technical results we use in previous chapters. Appendix A contains a proof of the following statement : for a diffeomorphism ϕ which satisfies the STC there is a constant L such that any δ-trajectory of ϕ with small δ is $L\delta$-traced by a real trajectory. In Appendix B we investigate attractors with the STC on the boundary. The structure of the boundary of the attractor in this case is described ; it is shown, for example, that the boundary is an attractor itself.

In Appendix C we study families of pseudotrajectories generated by numerical methods. We prove a theorem obtained by R.Corless and the author. It shows that for any diffeomorphism ϕ satisfying the STC there exist numerical methods of arbitrary accuracy such that ϕ has trajectories which are not weakly traced by approximate trajectories obtained using these methods.

We usually do not give in this book any special references to statements included in basic university courses of mathematics. The list of references is far from being complete. It contains only those books and research papers which are directly mentioned in the text.

Many conversations with colleagues were very important for the author during the preparation of this book. Special thanks are to D.Anosov, V.Arnold, R.Corless, Yu.Il'yashenko, O.Ivanov, V.Pliss, V.Pogonysheva, R.Russell, G.Sell, T.Eirola and to Zhang Zhi-fen.

Contents

List of Main Symbols

$\mathbf{R}$ – the real line;

$\mathbf{Z}$ – the set of integers;

$\mathbf{R}^n$ – the Euclidean n-space;

For a set A : $\overline{A}$ – the closure, Int$A$ – the interior, diam$A$ – the diameter, ∂A – the boundary, $N_\delta(A)$ – the δ-neighborhood;

For a topological space X : $\emptyset$ – the empty set, X^* – the set of compact subsets;

For a manifold M : dim$M$ – the dimension, id,id$_M$ – the identity map, T_xM – the tangent space at x, TM – the tangent bundle;

$Z(M)$ – the space of dynamical systems with the C^0- topology;

Diff$^r(M)$ – the space of C^r-diffeomorphisms with the C^r-topology;

CLD(M) – the closure of Diff$^1(M)$ in $Z(M)$;

$Df(p)$ – the derivative of a map f at p;

For a dynamical system ϕ : Fix(ϕ) – the set of fixed points, Per(ϕ) – the set of periodic points, $\Omega(\phi)$ – the nonwandering set, WPer(ϕ) – the set of weakly periodic points, W$\Omega(\phi)$ – the set of weakly nonwandering points, CR(ϕ) – the chain-recurrent set;

For a dynamical system ϕ and for a point x : $O(x, \phi)$ – the trajectory, $Q(x, \phi)$ – the prolongation with respect to the initial point, $P(x, \phi)$ – the prolongation with respect to the system, $R(x, \phi)$ – the chain prolongation, $Q^\omega(x, \phi), P^\omega(x, \phi), R^\omega(x, \phi)$ – the limit prolongations;

$\omega(G, \phi)$ – the ω-limit set of $G \subset M$ in $\phi \in Z(M)$;

$W^s(x)$ – the stable manifold of x;

$W^u(x)$ – the unstable manifold of x;

$D(\eta, p)$ – a smooth disc of radius η centered at p;

For a linear map A

$$\| A \| = \sup_{|x|=1} | Ax |;$$

card$K$ – the cardinality of a finite set K.

0. Definitions and Preliminary Results

0.1 Spaces of Dynamical Systems

In this book we study dynamical systems on closed smooth manifolds. Some results we prove have analogues for continuous dynamical systems on metric compact sets but we don't pay attention to possible generalizations of this kind.

This section is mostly devoted to fix the language and basic notation. Prerequisites for reading this book are basic courses on Dynamical Systems (see [Pa4, Pi8] for example) and on Differentiable Manifolds ([Hi1, Mu2] for example).

Throughout the book M is a C^∞-smooth closed (that is compact and without boundary) manifold of dimension n. We fix a Riemannian metric d on M. We denote by T_xM the tangent space of M at $x \in M$, and by TM the tangent bundle of M. For $v \in T_xM$ we denote by $|v|$ the norm generated on T_xM by d. With fixed Riemannian metric d for $x \in M$ we can define the exponential map exp_x being a diffeomorphism of class C^∞ of a neighborhood of 0 at T_xM onto a neighborhood of x in M. As M is compact there is $r > 0$ such that for any $x \in M$ the map exp_x is a diffeomorphism of $\{v \in T_xM : |v| < r\}$ onto its image.

We study discrete dynamical systems generated by homeomorphisms $\phi : M \to M$. We do not distinguish between a homeomorphism ϕ and the dynamical system it generates.

Let us denote by

$$O(x, \phi) = \{\phi^k(x) : k \in \mathbf{Z}\}$$

the trajectory (orbit) of a point $x \in M$ in a dynamical system ϕ. Sometimes we write $O(\phi)$ (if it is not important to mark the initial point) or $O(x)$ (if we work with a fixed dynamical system) instead of $O(x, \phi)$.

We denote

$$O^+(x, \phi) = \{\phi^k(x) : k \in \mathbf{Z}, k \geq 0\},$$
$$O^-(x, \phi) = \{\phi^k(x) : k \in \mathbf{Z}, k \leq 0\}.$$

We also use the following notation. If $x \in M$, $k_1, k_2 \in \mathbf{Z}, k_1 \leq k_2$ we write

$$O_{k_1}^{k_2}(x, \phi) = \{\phi^k(x) : k_1 \leq k \leq k_2\}.$$

In this case $k_1 = -\infty$ or $k_2 = +\infty$ are admissible, so that

$$O_{-\infty}^{k_2}(x, \phi) = \{\phi^k(x) : -\infty < k \leq k_2\},$$

and $O^-(x,\phi) = O^0_{-\infty}(x,\phi)$, for example. We also sometimes write $O^+(\phi), O^+(x)$ instead of $O^+(x,\phi)$. We use the same notation for trajectories of sets: if $X \subset M$ then we write

$$O(X,\phi) = \{\phi^k(X) : k \in \mathbf{Z}\},$$

and so on.

As usually we say that a point $x \in M$ is a *periodic point* of period m for a dynamical system ϕ if

$$\phi^m(x) = x; \phi^k(x) \neq x \text{ for } 0 < k \leq m-1$$

(if $\phi(x) = x$ we say that x is a fixed point). We denote by $\mathrm{Per}(\phi)$ the set of periodic points of ϕ, and by $\mathrm{Fix}(\phi)$ the set of fixed points of ϕ.

We say that a point $x \in M$ is a *nonwandering point* of ϕ if given a neighborhood U of x and a number $m_0 > 0$ we can find a number $m \in \mathbf{Z}, |m| \geq m_0$ such that

$$\phi^m(U) \cap U \neq \emptyset.$$

Denote by $\Omega(\phi)$ the set of nonwandering points of ϕ.

For a point $x \in M$ and for a dynamical system ϕ we define α_x, the α-limit set of $O(x,\phi)$, and ω_x, the ω-limit set of $O(x,\phi)$, as follows:

$$\alpha_x(\phi) = \{\lim_{k\to\infty} \phi^{t_k}(x) : \lim_{k\to\infty} t_k \to -\infty\},$$

$$\omega_x(\phi) = \{\lim_{k\to\infty} \phi^{t_k}(x) : \lim_{k\to\infty} t_k \to +\infty\}.$$

Sometimes we write α_x, ω_x instead of $\alpha_x(\phi), \omega_x(\phi)$. The following statement is standard (see [Pi8], for example).

Lemma 0.1.1.
(a) the set $\Omega(\phi)$ is compact and ϕ-invariant;
(b) $x \in \Omega(\phi)$ if and only if there exist sequences

$$x_k \to x,\ t_k \to +\infty \text{ as } k \to \infty$$

such that

$$\phi^{t_k}(x_k) \to x \text{ as } k \to \infty;$$

(c) for any $x \in M$ we have $\alpha_x(\phi) \cup \omega_x(\phi) \subset \Omega(\phi)$.

Take two discrete dynamical systems ϕ, ψ on M and let

$$\rho_0(\phi,\psi) = \max_{x\in M}(d(\phi(x),\psi(x)), d(\phi^{-1}(x),\psi^{-1}(x))).$$

It is easy to see that ρ_0 is a metric on the space of dynamical systems. The main object in this book is the space $Z(M)$ of continuous discrete dynamical systems on M with the C^0-topology induced by the metric ρ_0. Standard considerations show that $(Z(M),\rho_0)$ is a complete metric space.

For a set $A \subset M$, for a dynamical system $\phi \in Z(M)$, and for a number $\epsilon > 0$ we denote

$$N_\epsilon(A) = \{x \in M : d(x, A) < \epsilon\}$$

(here as usually $d(x, A) = \inf_{y \in A} d(x, y)$),

$$N_\epsilon(\phi) = \{\psi \in Z(M) : \rho_0(\phi, \psi) < \epsilon\}.$$

We consider not only continuous but also differentiable dynamical systems on M generated by diffeomorphisms of M. To introduce the C^r-topology on the space of diffeomorphisms of class $C^r, 1 \leq r < +\infty$, we can proceed as follows.

Consider a smooth map $f : M \to N$, where M, N are smooth manifolds.

The map f is said to be an *immersion* if the derivative

$$Df(p) : T_p M \to T_{f(p)} N$$

is injective for all $p \in M$.

The map f is said to be an *embedding* if f is an injective immersion which has a continuous inverse $f^{-1} : f(M) \subset N \to M$.

The classical Whitney's Theorem [Hi1] states that if $\dim M = n$ then there exists an embedding

$$f : M \to \mathbf{R}^{2n+1}. \tag{0.1}$$

Fix a finite covering of M by open sets $V_1, \ldots, V_m$ such that each $\overline{V_i}$ is contained in the domain of a local chart (ξ_j, U_j) of M.

Consider a smooth map $\chi : \mathbf{R}^n \to \mathbf{R}^q$, let $x = (x_1, \ldots, x_n)$ be coordinates of $\mathbf{R}^n$. As usually we denote for $p = (p_1, \ldots, p_n)$ with $p_i \in \mathbf{Z}, p_i \geq 0$

$$\frac{\partial^p \chi}{\partial x^p} = \frac{\partial^p \chi}{\partial x_1^{p_1} \ldots \partial x_n^{p_n}}, |p| = p_1 + \ldots + p_n.$$

Take two diffeomorphisms ϕ, ψ of M. Take $i \in \{1, \ldots, m\}$, suppose that $\overline{V_i} \subset U_j$, and let

$$\tilde{V}_i = \xi_j(V_i) \subset \mathbf{R}^n,$$

$$\tilde{\phi}_i = f \circ \phi \circ \xi_j^{-1}, \tilde{\psi}_i = f \circ \psi \circ \xi_j^{-1}$$

(here f is the embedding (0.1)). If ϕ, ψ are diffeomorphisms of class $C^r, 1 \leq r < +\infty$, we can introduce the number

$$\rho_r(\phi, \psi) = \max_{1 \leq i \leq m} \sup_{x \in \tilde{V}_i} \sum_{0 \leq |p| \leq r} \| \frac{\partial^p (\tilde{\phi}_i - \tilde{\psi}_i)}{\partial x^p}(x) \|$$

(as everywhere in this book for a linear operator A we consider the operator norm

$$\|A\| = \sup_{|y|=1} |Ay|.)$$

It is easy to see that ρ_r is a metric on the space of diffeomorphisms of class C^r of M. This metric induces the C^r-topology (as M is compact the topology is independent on the choice of $V_1, \dots, V_m$). We denote by $\mathrm{Diff}^r(M)$ the corresponding topological space. It is evident that for any $r \geq 1$ the topology of $\mathrm{Diff}^r(M)$ is not coarser than the topology on the space of C^r-diffeomorphisms induced by the topology of $Z(M)$.

We denote by $\mathrm{CLD}(M)$ the closure of $\mathrm{Diff}^1(M)$ in $Z(M)$. We shall use below the following important statement obtained independently by J.Munkres [Mu1] and by J.Whitehead [Wh].

Theorem 0.1.1. *If* $\dim M \leq 3$ *then* $Z(M) = CLD(M)$.

Note that it is shown in [Mu1] that if $\dim M > 3$ then there exist homeomorphisms which are not C^0-approximated by diffeomorphisms.

We do not discuss in this book analogues for flows of the results we describe. Nevertheless sometimes we work with flows - for example we apply techniques of flows to prove variants of the C^0-closing Lemma (Lemmas 0.3.1,0.3.3). Besides, it is very convenient to visualize some constructions using flows (see Sect. 0.1).

So we describe main notation we use below. We consider vector fields X which are Lipschitz on M in the following standard sense. Denote by π the projection $TM \to M$, that is for $(x,v) \in TM$ with $v \in T_xM$ we have $\pi(x,v) = x$. As usually we define a vector field X on M as a map $X : M \to TM$ such that $\pi \circ X(x) = x$ for any x. That means that we can write $X(p) = (p, X_p)$. Let $\tilde{d}$ be a metric on TM generated by d. We say that X is Lipschitz on M if there is a constant $L > 0$ such that for any $p_1, p_2 \in M$ we have

$$\tilde{d}(X(p_1), X(p_2)) \leq Ld(p_1, p_2).$$

It is easy to show that if a vector field X on M is of class C^1 then X is Lipschitz on M.

It is well-known that if a vector field X is Lipschitz then the corresponding system of differential equations

$$\frac{dx}{dt} = X(x)$$

generates a flow $\Phi : \mathbf{R} \times M \to M$ with $\Phi(0,x) = x$.

For two flows Φ_1, Φ_2 we let

$$\rho_0(\Phi_1, \Phi_2) = \max_{p \in M, t \in [-1,1]} d(\Phi_1(t,p), \Phi_2(t,p)).$$

It is easy to see that ρ_0 is a metric on the space of flows. Below we say that the topology induced by ρ_0 on the space of flows is the standard C^0-topology.

0.2 The Space M^*

We denote everywhere in this book by X^* the set of compact subsets of a topological space X. In this section we mostly pay attention to M^* - the set of compact subsets of our manifold M.

We begin with description of the standard *Hausdorff metric* R on M^*. Take $A, B \in M^*$ and let

$$r(A, B) = \max_{x \in A} d(x, B).$$

Take $A, B \in M^*$ and if $A, B \neq \emptyset$ let

$$R(A, B) = \max(r(A, B), r(B, A)).$$

Let for any $A \in M^*, A \neq \emptyset$

$$R(\emptyset, A) = \text{diam} M = \max_{x,y \in M} d(x, y),$$

and let

$$R(\emptyset, \emptyset) = 0.$$

It is easy to show that R is a metric on M^*. It is evident that for any A we have $R(A, A) = 0$ and that $R(A, B) = 0$ implies $A = B$. To prove the triangle inequality take $A, B, C \in M^*$. Consider $x \in A, y \in B$, then

$$d(x, C) \leq d(x, y) + d(y, C) \leq d(x, y) + R(B, C),$$

hence

$$d(x, C) \leq \min_{y \in B} d(x, y) + R(B, C) = d(x, B) + R(B, C),$$

and

$$d(x, C) \leq R(A, B) + R(B, C).$$

Therefore,

$$r(A, C) \leq R(A, B) + R(B, C).$$

The same reasons show that

$$r(C, A) \leq R(A, B) + R(B, C),$$

so we obtain that

$$R(A, C) \leq R(A, B) + R(B, C).$$

It follows that R is a metric in M^*. It is shown in [K] that (M^*, R) is a complete metric space. Later if we write M^* we have in mind the space (M^*, R).

We work in this book also with the following metrics R_0, R_1, R_2 on M^* (we use for R_0 the original notation from the paper [Pi3] in which this metric was introduced; R_2 was introduced in [Po1] and was denoted R_1 there). Take $A, B \in M^*$ and let

$$R_0(A, B) = \max(R(A, B), R(\overline{M \setminus A}, \overline{M \setminus B})),$$

$$R_1(A, B) = \max(R(A, B), R(\overline{\text{IntA}}, \overline{\text{IntB}})),$$

$$R_2(A, B) = \max(R_0(A, B), R_1(A, B)).$$

It is evident that for $S = R_0, R_1, R_2$ we have $S(A, B) \geq 0, S(A, A) = 0$, and that $S(A, B) = 0$ implies $A = B$.

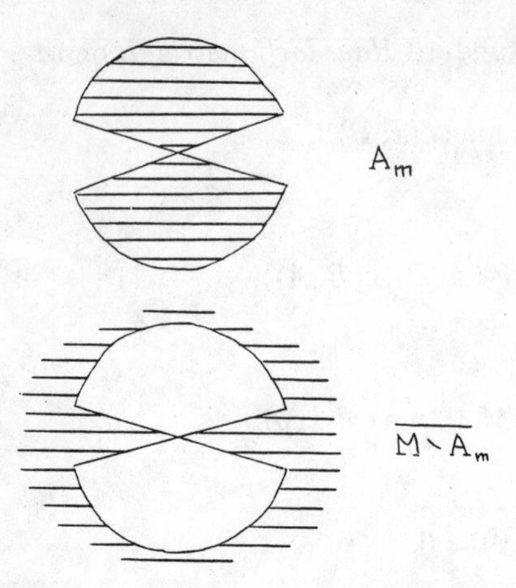

Figure 0.1

If $\rho_1, \rho_2 : M^* \times M^* \to \mathbf{R}$ satisfy the triangle inequality and $\rho(A,B) = \max(\rho_1(A,B), \rho_2(A,B))$ then ρ also satisfies the triangle inequality. Indeed, take $A, B, C \in M^*$. If $\rho(A,C) = \rho_1(A,C)$ then

$$\rho(A,C) \le \rho_1(A,B) + \rho_1(B,C) \le$$

$$\max(\rho_1(A,B), \rho_2(A,B)) + \max(\rho_1(B,C), \rho_2(B,C)) = \rho(A,B) + \rho(B,C).$$

If $\rho(A,C) = \rho_2(A,C)$ the reasons are the same. Hence, R, R_0, R_1, R_2 are metrics on M^*.

Let us show that R, R_0, R_1, R_2 induce different topologies on M^*. To show this, consider $M = S^2$ and take a coordinate neighborhood being homeomorphic to $\mathbf{R}^2$ with coordinates (x, y).

Let $A = \{(x,y) : x^2 + y^2 \le 1\}$ and for $m \ge 1$

$$A_m = A \bigcap \{(x,y) : |x| \ge \frac{|y|}{m}\}$$

(see Fig. 0.1). It is evident that

$$\lim_{m\to\infty} R(A, A_m) = 0, R_0(A, A_m) \equiv 1,$$

$$\lim_{m\to\infty} R_1(A, A_m) = 0.$$

Now let

$$A = \{(x,y) : x^2 + y^2 \le 1\} \bigcup \{(x,y) : 1 \le x \le 2, y = 0\},$$

and for $m \ge 1$

$$A_m = A \bigcup \{(x,y) : 0 \le x \le 2, \frac{x-2}{m} \le y \le \frac{2-x}{m}\}$$

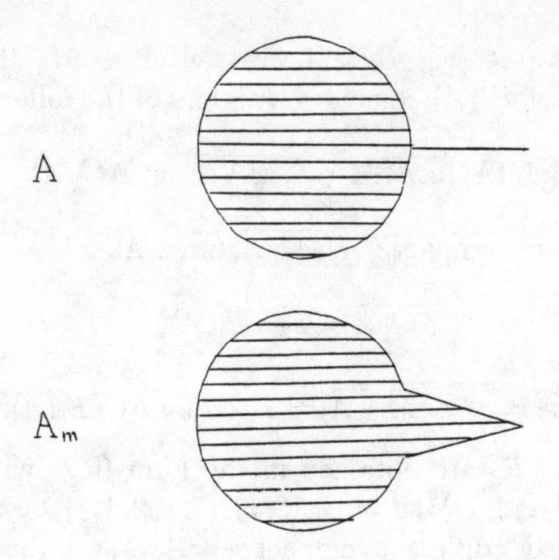

Figure 0.2

(see Fig. 0.2). It is evident that

$$\lim_{m\to\infty} R(A, A_m) = 0, \lim_{m\to\infty} R_0(A, A_m) = 0,$$

$$R_1(A, A_m) \equiv 1.$$

Now we describe a general process of constructing metrics on M^*. We are going to show that the metrics R, R_0, R_1, R_2 form the complete list of metrics given by this process.

Let for a set $A \subset M$

$$F_1(A) = M \setminus A, F_2(A) = \overline{A}.$$

Consider the following set of finite sequences

$$J = \{(i_1, \ldots, i_m) : i_k \in \{1, 2\}\},$$

and let

$$F^{\emptyset}(A) = A,$$

for $j = \{i_1, \ldots, i_m\} \in J$ let

$$F^j(A) = F_{i_1} \circ F_{i_2} \circ \ldots \circ F_{i_m}(A).$$

Now take a finite subset $\tilde{J}$ of J which has the properties:

(a) for any $j \in \tilde{J}, A \in M^*$ we have $F^j(A) \in M^*$;

(b) there exists $j \in \tilde{J}$ such that $R(F^j(.), F^j(.))$ is a metric on M^*.

Define for $A, B \in M^*$

$$R_{\tilde{J}}(A, B) = \max_{j\in\tilde{J}}(R(F^j(A), F^j(B))). \tag{0.2}$$

It follows from our previous considerations that any $R_{\tilde{J}}$ is a metric on M^*.

It is well-known (see [K]) that if we take $A \in M^*$ then for any $j = (i_1, \ldots, i_m) \in J$ the set $F^j(A)$ coincides with one of the following sets:

$$A, \mathrm{IntA}, \overline{\mathrm{IntA}}, M \setminus A, \overline{M \setminus A}, \mathrm{Int}(M \setminus A) \tag{0.3}$$

Only 3 sets in (0.3) are compact : $A, \overline{M \setminus A}, \overline{\mathrm{IntA}}$. As

$$\overline{M \setminus A} = F_2 \circ F_1(A),$$

$$\overline{\mathrm{IntA}} = \overline{M \setminus (\overline{M \setminus A})} = F_2 \circ F_1 \circ F_2 \circ F_1(A),$$

the metrics R, R_0, R_1, R_2 are obtained in the form (0.2) with $\tilde{J} = \{\emptyset\}, \tilde{J} = \{\emptyset, (2,1)\}, \tilde{J} = \{\emptyset, (2,1,2,1,)\}, \tilde{J} = \{\emptyset, (2,1), (2,1,2,1,)\}$, respectively.

Evidently there exist different compact sets A, B such that $\overline{M \setminus A} = \overline{M \setminus B}$ and $\overline{\mathrm{IntA}} = \overline{\mathrm{IntB}}$, hence

$$R(\overline{M \setminus A}, \overline{M \setminus B}), R(\overline{\mathrm{IntA}}, \overline{\mathrm{IntB}}),$$

$$\max(R(\overline{M \setminus A}, \overline{M \setminus B}), R(\overline{\mathrm{IntA}}, \overline{\mathrm{IntB}}))$$

are not metrics on M^*. This proves that $\{R, R_0, R_1, R_2\}$ is the complete list of metrics obtained by the described process.

We shall consider also M^{**} – the set of all compact subsets of M^* (we remind that here M^* means (M^*, R)). We denote by $\bar{R}$ the Hausdorff metric on M^{**}.

Let now X be a topological space. A subset $A \subset X$ is called *residual* if A contains a countable intersection of open dense sets in X. If P is a property of elements of X we say that this property is *generic* if the set $\{x \in X : x \text{ satisfies } P\}$ is residual in X. Sometimes in this case we say that a generic element of X satisfies P.

The topological space X is called a Baire space if every residual set is dense in it. A classical theorem of Baire [K] says that every complete metric space is a Baire space. Hence, $Z(M)$ is a Baire space, and $\mathrm{CLD}(M)$ is also a Baire space.

Now we are going to prove a result which is very useful to establish genericity of some properties of dynamical systems in $Z(M)$. Let us begin with the following definition.

Let X be a topological space, and let (N, ρ) be a compact metric space. A map

$$\psi : X \to N^*$$

is called *upper semi-continuous* (respectively, *lower semi-continuous*) if for every $x \in X$ and $\delta > 0$ there is a neighborhood $W(x)$ of x in X such that for any $y \in W(x)$ we have

$$\psi(y) \subset N_\delta(\psi(x))$$

(respectively,

$$\psi(x) \subset N_\delta(\psi(y)) \text{).}$$

As previously, $N_\delta(\psi(x))$ is the δ-neighborhood of $\psi(x)$.

Let R be the Hausdorff metric on N^*. Fix $\delta > 0$. We say that the map ψ is δ-continuous at $x \in X$ if x has a neighborhood $W(x)$ in X such that for any $x', x'' \in W(x)$ we have

$$R(\psi(x'), \psi(x'')) < \delta.$$

Clearly ψ is continuous at $x \in X$ if it is δ-continuous at x for any $\delta > 0$. Let V_δ be the set of points of δ- continuity of ψ.

Lemma 0.2.1 [Ta1]. *If $\psi : X \to N^*$ is upper semi-continuous or lower semi-continuous then for any $\delta > 0$ the set V_δ is open and dense in X.*

Proof. . We consider the case of ψ being upper semi-continuous, the case of ψ being lower semi-continuous is treated analogously. It follows immediately from the definition that the set V_δ is open. Let us show that this set is dense.

Fix an open covering $U_1, \ldots, U_k$ of N such that diam $U_i < \delta, i = 1, \ldots, k$. Let $K = \{1, \ldots, k\}$ (we consider K as a discrete topological space), and let K^* be the set of subsets of K (we are following our notation here taking into account that any subset of K is compact).

Take $L \in K^*$ and consider the subset N_L of N^* : $A \in N^*$ is in N_L if and only if

(a) $A \cap U_i \neq \emptyset$ for any $i \in L$;

(b) $A \subset \bigcup_{i \in L} U_i$.

It is evident that for any L, N_L is open in N^*.

Now fix an open subset W of X. Define

$$B_W = \{L \in K^* : \text{ there is } x \in W \text{ with } \psi(x) \in N_L\}.$$

As K^* is finite and has a partial order (by the inclusion relation for subsets of K) we can find a minimal element $L_0 \in B_W$. This means that there is no $L \in B_W$ with $L \subset L_0, L \neq L_0$.

Let $x_0 \in W$ be such that $\psi(x_0) \in N_{L_0}$. We claim that ψ is δ-continuous at x_0. Indeed, as ψ is upper semi-continuous we can find a neighborhood $W(x_0)$ such that for any $x' \in W(x_0)$ we have

$$\psi(x') \subset \bigcup_{i \in L_0} U_i.$$

As L_0 is minimal we see that for any $x' \in W(x_0)$

$$\psi(x') \cap U_i \neq \emptyset, i \in L_0.$$

Take $x', x'' \in W(x_0)$ and consider $y' \in \psi(x')$. There is $i \in L_0$ such that $y' \in U_i$. Find $y'' \in \psi(x'')$ such that $y'' \in U_i$. It follows from diam $U_i < \delta$ that $\rho(y', y'') < \delta$. This evidently implies

$$R(\psi(x'), \psi(x'')) < \delta.$$

So, V_δ is dense. This completes the proof.

Corollary 0.1 *If $\psi : X \to N^*$ is upper semi-continuous or lower semi-continuous then the set of continuity points of ψ is residual in X.*

0.3 The C^0-closing Lemma

In the theory of structural stability a very important result was proved by C.Pugh [Pu]: if $\phi \in \text{Diff}^1(M)$ and $x_0 \in \Omega(\phi)$ then given $\epsilon > 0$ there is a diffeomorphism ψ such that $\rho_1(\phi, \psi) < \epsilon$ and $x_0 \in \text{Per}(\psi)$. This statement is usually called the C^1-closing Lemma. All known proofs of the C^1-closing Lemma are very complicated.

In this book we need analogues of this result for C^0-small perturbations instead of C^1-small ones. These statements belong to "folklore" and have simple proofs comparing to proofs of the C^1-closing Lemma. The main result we need is the following one.

Lemma 0.3.1 (the C^0-closing Lemma). *Let $\phi \in Z(M)$. Assume that Γ is a smooth segment embedded in M with ends p, q. Given a neighborhood U of Γ there is a system $\psi \in Z(M)$ such that*
(a) $\psi(p) = \phi(q)$;
(b) $\psi(x) = \phi(x)$ for $x \notin U$;
(c) $\rho(\phi, \psi) \leq \max(\text{ diam } U, \text{diam}\phi(U))$.

Proof. As Γ is a smooth embedded segment we can introduce coordinates $\xi \in \mathbf{R}^{n-1}, \eta \in \mathbf{R}^1$ in a neighborhood of Γ so that for some $a > 0$

$$\Gamma = \{(\xi, \eta) : \xi = 0, 0 \leq \eta \leq a\}, p = (0,0), q = (0, a).$$

Find $\delta > 0$ with the following property:

$$V = \{\xi : |\xi| \leq \delta\} \times \{\eta : -\delta \leq \eta \leq a + \delta\} \subset U.$$

Fix two functions $\alpha : \mathbf{R}^{n-1} \to \mathbf{R}, \beta : \mathbf{R} \to \mathbf{R}$ of class C^∞ and such that

$$\begin{aligned}
\alpha(0) &= 1; \\
0 < \alpha(\xi) &< 1 && \text{for } 0 < |\xi| < \delta; \\
\alpha(\xi) &= 0 && \text{for } |\xi| \geq \delta; \\
\beta(\eta) &= 1 && \text{for } 0 \leq \eta \leq a; \\
0 < \beta(\eta) &< 1 && \text{for } \eta \in (-\delta, 0) \cup (a, a+\delta); \\
\beta(\eta) &= 0 && \text{for } \eta \in (-\infty, -\delta] \cup [a+\delta, +\infty).
\end{aligned}$$

Using of functions of this sort is standard, we leave it to the reader to construct them. Let $\Xi(t, \xi, \eta), H(t, \xi, \eta)$ be the solution of the system of differential equations

$$\frac{d\xi}{dt} = 0, \frac{d\eta}{dt} = a\alpha(\xi)\beta(\eta) \tag{0.4}$$

with initial values $(0, \xi, \eta)$. Define $h : M \to M$ by

$$h(\xi, \eta) = (\Xi(1, \xi, \eta), H(1, \xi, \eta)) \text{ for } x = (\xi, \eta) \in V,$$

$$h(x) = x \text{ for } x \in M \setminus V.$$

As V is evidently invariant with respect to the flow (Ξ, H) , h is a diffeomorphism of M. It is easy to see that $\xi(t) = 0, \eta(t) = at$ is a solution of (0.4) for $t \in [0,1]$. Hence, $h(p) = q$. Take now $\psi = \phi \circ h$. Evidently $\psi(p) = \phi(q)$, and $\phi(x) = \psi(x)$ for $x \in M \setminus V$. This proves the statements (a) and (b) of our lemma. As for $x \in U$ both $\phi(x), \psi(x) \in \phi(U)$, and for $y \in \phi(U)$ both $\phi^{-1}(y), \psi^{-1}(y) \in U$ we obtain that the third statement is also true.

It will be convenient for us to use the following technical result based on Lemma 0.3.1.

Lemma 0.3.2. *Consider a system $\phi \in Z(M)$, and let ϕ_m be a sequence of systems with*

$$\lim_{m\to\infty} \rho_0(\phi, \phi_m) = 0.$$

Assume that for sequences of points x_m, y_m and for a sequence of numbers $k(m) > 0$ we have

$$\lim_{m\to\infty} x_m = x^*, \lim_{m\to\infty} y_m = x^*,$$

$$\lim_{m\to\infty} \phi_m^{k(m)}(y_m) = y^*, x^* \neq y^*.$$

Then there exists a sequence of systems ψ_m and a sequence of numbers $\lambda(m) > 0$ such that

(a) $\lim_{m\to\infty} \rho_0(\psi_m, \phi) = 0$;
(b) $\psi_m^{\lambda(m)}(x_m) = \phi_m^{k(m)}(y_m)$.

Proof. Fix arbitrary $\epsilon > 0$. Find $\delta > 0$ such that $\delta < \epsilon$ and for any $x_1, x_2 \in M$ with $d(x_1, x_2) < \delta$ we have

$$d(\phi(x_1), \phi(x_2)) < \frac{\epsilon}{3}.$$

Take δ so small that $2\delta < d(x^*, y^*)$. There is m_0 such that for $m \geq m_0$

$$\rho_0(\phi, \phi_m) < \frac{\epsilon}{3}. \tag{0.5}$$

If for $x_1, x_2 \in M$ we have $d(x_1, x_2) < \delta$ then

$$d(\phi_m(x_1), \phi_m(x_2)) \leq d(\phi_m(x_1), \phi(x_1)) +$$

$$+ d(\phi(x_1), \phi(x_2)) + d(\phi(x_2), \phi_m(x_2)) < \epsilon$$

for $m \geq m_0$. Let

$$Y_m = \{y_m, \phi_m(y_m), \ldots, \phi_m^{k(m)}(y_m)\}.$$

Find $m_1 \geq m_0$ such that for $m \geq m_1$

$$\min_{y \in Y_m} d(x_m, y) < \frac{\delta}{2}, d(\phi_m^{k(m)}(y_m), y^*) < \frac{\delta}{2}, d(x_m, x^*) < \frac{\delta}{2}. \tag{0.6}$$

Fix $m \geq m_1$ and find a point $\tilde{y}_m = \phi_m^\mu(y_m) \in Y_m$ with

$$d(x_m, \tilde{y}_m) = \min_{y \in Y_m} d(x_m, y). \tag{0.7}$$

It follows from (0.6) and from the choice of δ that if (0.7) is satisfied then $\mu < k(m)$. Take $\tilde{y}_m$ (and μ) such that

$$d(x_m, \phi_m^k(y_m)) > d(x_m, \tilde{y}_m) \text{ for } \mu < k \leq k(m).$$

Let Γ be a geodesic segment joining $x_m, \tilde{y}_m$ and such that diam$\Gamma < \delta$. As $\phi_m^k(y_m) \notin \Gamma$ for $\mu < k \leq k(m)$ there is a neighborhood U of Γ having the properties:

$$\text{diam}U < \delta, \phi_m^k(y_m) \notin U \text{ for } \mu < k \leq k(m).$$

By the choice of m, δ we have

$$\text{diam}U < \epsilon, \text{diam}\phi_m(U) < \epsilon. \tag{0.8}$$

Apply Lemma 0.3.1 to find a system $\psi_m \in Z(M)$ such that $\psi_m(x_m) = \phi_m(\tilde{y}_m), \psi_m(x) = \phi_m(x)$ for $x \notin U$, and $\rho_0(\phi, \psi_m) < 2\epsilon$ (we use (0.5),(0.8) to obtain the last inequality).

It is easy to understand that

$$\psi_m(x_m) = \phi_m(\tilde{y}_m) = \phi_m^{\mu+1}(y_m), \ldots, \psi_m^{\lambda(m)}(x_m) = \phi_m^{k(m)}(y_m)$$

where $\lambda(m) = k(m) - \mu$. As ϵ is arbitrary this proves our lemma.

We prove below one more "C^0-closing" statement obtained by Z.Nitecki and M.Shub in [Ni3]. Note that in contrast with Lemmas 0.3.1, 0.3.2 this result needs a dimension restriction on M.

Lemma 0.3.3. *Assume that dimM ≥ 2. Consider a finite collection*

$$\{(p_i, q_i) \in M \times M; \ i = 1, \ldots, k\}$$

such that

(a) $p_i \neq p_j, q_i \neq q_j$ for $1 \leq i < j \leq k$;
(b) for a small positive δ $d(p_i, q_i) < \delta, i = 1, \ldots, k$.
Then there exists $f \in \text{Diff}^1(M)$ with the following properties:
(a) $\rho_0(f, id) < 2\delta$;
(b) $f(p_i) = q_i, i = 1, \ldots, k$.

Proof. Take S^1 with coordinate $\theta \in [0, 1)$. Consider a system of differential equations on $M \times S^1$ with coordinates (p, θ):

$$\dot{p} = 0, \dot{\theta} = 1.$$

Its vector field is $X = (0, 1)$ and its flow is given by $\Phi(t, p, \theta) = (p, \theta + t(mod 1))$. Evidently $\Phi(1, p, \theta)$ takes $M \times 0$ to itself and induces the identity map there.

Given the points $p_i, q_i \in M$ we consider the points $(p_i, \frac{1}{4}), (q_i, \frac{3}{4})$ in $M \times S^1$. We take for each i a curve $\gamma_i(t)$ in $M, 0 \leq t \leq \frac{1}{2}$, of constant speed, joining p_i to q_i, of length less than δ. We can change parameter t on $\gamma_i(t)$ so that

$$\dot{\gamma}_i(0) = \dot{\gamma}_i(\frac{1}{2}) = 0, |\dot{\gamma}_i(t)| < 2\delta. \tag{0.9}$$

Consider the curves g_i given by

$$g_i(t) = (\gamma_i(t), \frac{1}{4} + t), 0 \leq t \leq \frac{1}{2}.$$

As $\dim(M \times S^1) \geq 3$ it follows from the Transversality Theorem (see [Hi1]) that we can slightly perturb the curves $\gamma_i(t)$ so that (0.9) holds and

$$g_i(t) \neq g_j(t) \text{ for } 0 \leq t \leq \frac{1}{2}, 1 \leq i < j \leq k.$$

Hence we can find tubular neighborhoods N_i of g_i in $M \times S^1$ with

$$\bar{N}_i \cap \bar{N}_j = \emptyset, 1 \leq i < j \leq k.$$

Take the tubular neighborhood N_i of g_i, extend g_i beyond its endpoints, and then extend N_i to be a tubular neighborhood of the extended curve (let this extended neighborhood be N_i^*). This can be done so that

$$\bar{N}_i^* \cap \bar{N}_j^* = \emptyset, 1 \leq i < j \leq k. \tag{0.10}$$

Let Y_i be the vector field $Y_i = (\dot{\gamma}_i(t), 1)$ on g_i (we can take $Y_i = (0,1)$ on the extended part of g_i). Extend Y_i to N_i^* making it constant along fibers. Now take a "bump" function Θ which is 1 on $g_i(t), 0 \leq t \leq \frac{1}{2}$, and 0 off N_i^*. We obtain the vector field

$$Z_i = X + \Theta(Y_i - X)$$

such that $Z_i = (\dot{\gamma}_i(t), 1) = \dot{g}_i(t)$ on $g_i(t), 0 \leq t \leq \frac{1}{4}$, hence $g_i(t)$ is an integral curve of Z_i, and

$$|Z_i - X| < 2\delta. \tag{0.11}$$

It follows from (0.10) that we can obtain a vector field Z which coincides with X on $M \setminus (N_1^* \cup \ldots \cup N_k^*)$, and with Z_i on $N_i^*, i = 1, \ldots, k$. Let $\Psi(t,p,\theta)$ be the flow of the corresponding system of differential equations. As $X = (0,1)$, and the inequalities (0.11) hold for $i = 1, \ldots, k$, we obtain from standard estimates for differential equations that for any $(p, \theta) \in M \times S^1$ we have

$$d(\Psi(t,p,\theta), \Phi(t,p,\theta)) < 2\delta$$

if $|t| \leq 1$. Let f be the time-one map of the flow Ψ. It follows from our construction that $f(p_i) = q_i, i = 1, \ldots, k$, and $\rho_0(f, id) < 2\delta$.

0.4 Hyperbolic Sets

The notion of a hyperbolic set is the basic notion in the theory of structural stability of smooth dynamical systems. Methods of this theory developed to investigate hyperbolic sets play also very important role in the theory of C^0-small perturbations of dynamical systems. Many results included in this book are based on the property of hyperbolicity. Here we give a brief survey (without any proofs) of the results we need.

Let us begin with definitions. Fix a diffeomorphism ϕ of class C^1. Consider a point $x \in M$ and numbers $C > 0, \lambda \in (0,1)$.

We say that the trajectory $O(x,\phi)$ is *hyperbolic* with hyperbolicity constants C, λ (and we write $x \in H(C,\lambda), O(x,\phi) \in H(C,\lambda)$ in this case) if for any $p \in O(x,\phi)$ there exist two linear subspaces E_p^s, E_p^u of T_pM such that:

(a) $E_p^s + E_p^u = T_pM$;

(b) $D\phi(p)E_p^s = E_{\phi(p)}^s$;

$D\phi(p)E_p^u = E_{\phi(p)}^u$;

(c) $|D\phi^k(p)v| \leq C\lambda^k|v|$ for $v \in E_p^s, k \geq 0$;

$|D\phi^k(p)v| \leq C\lambda^{-k}|v|$ for $v \in E_p^u, k \leq 0$.

Now we define a *hyperbolic set.* We say that a subset $I \subset M$ is a hyperbolic set of ϕ if:

(a) I is compact and ϕ-invariant;

(b) there exist constants $C > 0, \lambda \in (0,1)$ such that for any $x \in I$ we have $O(x,\phi) \in H(C,\lambda)$.

In this case we write $I \in H(C,\lambda)$. The reader is referred to Chap. 12 of the book [Pi8] for proofs of basic properties of hyperbolic sets. It is easy to see that if I is a hyperbolic set then for any $p \in I$

$$E_p^s \oplus E_p^u = T_pM$$

(here $\oplus$ denotes direct sum). So we can write for the restriction of the tangent bundle TM of M on I:

$$TM\,|_I = E^s \oplus E^u \tag{0.12}$$

where subbundles E^s, E^u are defined in the following natural way:

$$E^\sigma = \{(p, E_p^\sigma) : p \in I\}\ , \sigma = s, u.$$

We say below that (0.12) is the hyperbolic structure on I with hyperbolicity constants C, λ.

Let $I \in H(C,\lambda)$. We can define a metric d_L on M such that for the corresponding norm $|\ |_L$ the following holds: there exists $\mu \in (\lambda, 1)$ such that for $x \in I$

$$\begin{aligned} |D\phi(x)v|_L &\leq \mu|v|_L \quad \text{for } v \in E_x^s, \\ |D\phi^{-1}(x)v|_L &\leq \mu|v|_L \quad \text{for } v \in E_x^u. \end{aligned}$$

See, for example, Sect. 9 of Chap. 2 in [An1]. A metric d_L which has the described property is called a *Lyapunov metric* for the hyperbolic set I. With respect to this metric we can write

$$\|D\phi\,|_{E_x^s}\,\| \leq \mu, \|D\phi^{-1}\,|_{E_x^u}\,\| \leq \mu$$

for $x \in I$. In this case we say that (0.12) is the hyperbolic structure on I with hyperbolicity constant μ.

Now we describe geometric objects which play the crucial role in the theory of hyperbolic sets. Let $p \in I \in H(C, \lambda)$. Define the *stable* and the *unstable manifolds* of p:

$$W^s(p) = \{x \in M : \lim_{k\to\infty} d(\phi^k(x), \phi^k(p)) = 0\},$$

$$W^u(p) = \{x \in M : \lim_{k\to-\infty} d(\phi^k(x), \phi^k(p)) = 0\}.$$

Fix a point $p \in M$, an integer $k, 0 \leq k \leq n$, and a number $\eta > 0$. We say that $D(\eta, p)$ is a smooth (of class C^r) k-dimensional disc of radius η centered at p if we can write T_pM as $\mathbf{R}^k \times \mathbf{R}^{n-k}$ with coordinates y in $\mathbf{R}^k, z$ in $\mathbf{R}^{n-k}$ so that

$$D(\eta, p) = exp_p\{(y, f(y)) : y \in \mathbf{R}^k, |y| \leq \eta\}$$

where f is a C^r map from $\{y \in \mathbf{R}^k : |y| \leq \eta\}$ into $\mathbf{R}^{n-k}$, and $f(0) = 0$. Below we sometimes do not indicate the radius and the center for a disc D we work with.

The following result describes basic properties of stable and unstable manifolds. The reader can find a proof in [Pl1], Chap. 1.

Theorem 0.4.1 (The Stable Manifold Theorem). *Assume that* $\phi \in \text{Diff}^r(M), r \geq 1$ *and that* I *is a hyperbolic set of* ϕ. *Then there exists* $\epsilon_0 > 0$ *having the following properties. If* $p \in I, \dim E_p^s = l$, *then*

(a) there exist immersions b^s, b^u *of class* C^r:

$$b^s : \mathbf{R}^l \to M, b^s(0) = p, b^s(\mathbf{R}^l) = W^s(p);$$

$$b^u : \mathbf{R}^{n-l} \to M, b^u(0) = p, b^u(\mathbf{R}^{n-l}) = W^u(p);$$

(b) for any $\epsilon \in (0, \epsilon_0]$ *there exist smooth discs (of class* C^r*)* $W_\epsilon^s(p), W_\epsilon^u(p)$ *being subsets of*

$$W^s(p) \cap N_\epsilon(p), W^u(p) \cap N_\epsilon(p),$$

containing p *and such that*

(b.1) $T_pW_\epsilon^s(p) = E_p^s, T_pW_\epsilon^u(p) = E_p^u$;

(b.2) for $x \in N_\epsilon(p) \setminus W_\epsilon^s(p)$ *there is* $k_1 > 0$ *with*

$$d(\phi^{k_1}(x), \phi^{k_1}(p)) \geq \epsilon;$$

(b.3) for $x \in N_\epsilon(p) \setminus W_\epsilon^u(p)$ *there is* $k_2 < 0$ *with*

$$d(\phi^{k_2}(x), \phi^{k_2}(p) \geq \epsilon.$$

Now we define two basic classes of diffeomorphisms which are studied in the theory of structural stability: *Ω-stable diffeomorphisms* and *structurally stable diffeomorphisms.*

We say that $\phi \in \mathrm{Diff}^1(M)$ is Ω-stable if given $\epsilon > 0$ there is $\delta > 0$ such that for any $\psi \in \mathrm{Diff}^1(M)$ with $\rho_1(\phi, \psi) < \delta$ there is a homeomorphism $h : \Omega(\phi) \to \Omega(\psi)$ with

(a) $h \circ \phi \mid_{\Omega(\phi)} = \psi \circ h \mid_{\Omega(\phi)}$;

(b) $d(x, h(x)) < \epsilon$ for $x \in \Omega(\phi)$.

We say that $\phi \in \mathrm{Diff}^1(M)$ is structurally stable if given $\epsilon > 0$ there is $\delta > 0$ such that for any $\psi \in \mathrm{Diff}^1(M)$ with $\rho_1(\phi, \psi) < \delta$ there is a homeomorphism $h : M \to M$ with

(a) $h \circ \phi = \psi \circ h$;

(b) $d(x, h(x)) < \epsilon$ for $x \in M$.

S.Smale introduced in [Sm2] the following property of ϕ.

Axiom A:

(a) *$\Omega(\phi)$ is a hyperbolic set;*

(b) *the set* $\mathrm{Per}(\phi)$ *is dense in $\Omega(\phi)$.*

He showed that for a diffeomorphism ϕ which satisfies Axiom A we can describe the structure of the nonwandering set $\Omega(\phi)$ in the following way [Sm2].

Theorem 0.4.2 (The Spectral Decomposition Theorem). *If a diffeomorphism ϕ satisfies Axiom A then there exists a unique decomposition*

$$\Omega(\phi) = \Omega_1 \cup \ldots \cup \Omega_m, \tag{0.13}$$

here Ω_i are compact disjoint invariant sets, and each Ω_i contains a dense trajectory.

The sets $\Omega_1, \ldots, \Omega_m$ in the decomposition (0.13) are called *basic sets.* Define the following sets

$$W^s(\Omega_i) = \{x \in M : \lim_{k\to\infty} d(\phi^k(x), \Omega_i) = 0\},$$

$$W^u(\Omega_i) = \{x \in M : \lim_{k\to-\infty} d(\phi^k(x), \Omega_i) = 0\},$$

$i = 1, \ldots, m$. These sets are analogues of stable and unstable manifolds for individual hyperbolic trajectories.

S.Smale proved in [Sm2] the following

Theorem 0.4.3. *If a diffeomorphism ϕ satisfies Axiom A then*

$$M = \bigcup_{1\le i\le m} W^s(\Omega_i) = \bigcup_{1\le i\le m} W^u(\Omega_i).$$

Let us describe some properties of basic sets.

Theorem 0.4.4. *Let Ω_i be a basic set of a diffeomorphism ϕ which satisfies Axiom A. Then:*

(a) there is a neighborhood U of Ω_i such that

$$\bigcap_{k\in\mathbf{Z}} \phi^k(U) = \Omega_i$$

(this property is called the local maximality of Ω_i);
(b) $W^\sigma(\Omega_i) = \bigcup_{p\in\Omega_i} W^\sigma(p), \sigma = s, u$;
(c) for any $p \in \Omega_i$

$$W^\sigma(\Omega_i) \subset \overline{\bigcup_{r\in O(p,\phi)} W^\sigma(r)}, \sigma = s, u.$$

The statements (a),(b) were proved by M.Hirsch, J.Palis, C.Pugh and M.Shub [Hi2], the statement (c) follows from the density of Per(ϕ) in Ω_i and from the following important result of J.Palis (Theorem 0.4.5).

Let p be a hyperbolic periodic point of a diffeomorphism ϕ. As $O(p,\phi)$ is a hyperbolic set of ϕ we can apply Theorem 0.4.1 to find $\epsilon_0 > 0$ and a corresponding smooth disc

$$W^u_{\epsilon_0}(p) = exp_p\{(y, f(y)) : y \in \mathbf{R}^l \subset T_pM, |y| \leq \epsilon_0\}$$

in $W^u(p)$ with $f \in C^1, f(0) = 0$.

Take two discs D_1, D_2 given by

$$D_i = exp_p\{(y, f_i(y)) : y \in \mathbf{R}^l \subset T_pM, |y| \leq \epsilon_0\},$$

$i = 1, 2$, with $f_i \in C^1$, and let

$$\rho_1(D_1, D_2) = \sup_{|y|\leq\epsilon_0} (|f_1(y) - f_2(y)| + \|\frac{\partial f_1}{\partial y}(y) - \frac{\partial f_2}{\partial y}(y)\|).$$

Theorem 0.4.5 (the λ-Lemma) [Pal]. *Assume that N is a smooth disc having a point of transversal intersection with $W^s(p)$. Given $\epsilon > 0$, there exists $k(\epsilon)$ such that for $k \geq k(\epsilon)$ we can find functions $f_k \in C^1$,*

$$f_k : \{y \in \mathbf{R}^l \subset T_pM : |y| \leq \epsilon_0\} \to \{\mathbf{R}^{n-l} \subset T_pM\}$$

such that the discs

$$N_k = exp_p\{(y, f_k(y)) : y \in \mathbf{R}^l \subset T_pM, |y| \leq \epsilon_0\}$$

have the following properties :
(a) $N_k \subset \phi^k(N)$;
(b) $\rho_1(N_k, W^u_{\epsilon_0}(p)) < \epsilon$.

Now take two different basic sets Ω_i, Ω_j. We write $\Omega_i \to \Omega_j$ if

$$W^u(\Omega_i) \cap W^s(\Omega_j) \neq \emptyset.$$

We say that ϕ has a 1-cycle if there exists a basic set Ω_i such that

$$(W^u(\Omega_i) \cap W^s(\Omega_i)) \setminus \Omega_i \neq \emptyset.$$

We say that ϕ has a k-cycle, $k > 1$, if there exist k different basic sets $\Omega_{i_1}, \ldots, \Omega_{i_k}$ such that

$$\Omega_{i_1} \to \Omega_{i_2} \ldots \to \Omega_{i_k} \to \Omega_{i_1}.$$

We say that ϕ satisfies the *no-cycle condition* if ϕ has no k-cycles with $k \geq 1$.

Now we can describe necessary and sufficient conditions for a diffeomorphism ϕ to be Ω-stable.

Theorem 0.4.6 [Sm3, Pa5]. *A diffeomorphism ϕ is Ω-stable if and only if ϕ satisfies Axiom A and the no-cycle condition.*

Remark. S.Smale proved in [Sm3] that Axiom A and the no-cycle condition imply Ω-stability. J.Palis applied ideas of R.Mañé [Ma2] to establish the converse statement.

We say that a diffeomorphism ϕ satisfies the *geometric strong transversality condition* (geometric STC) if for any $p, q \in \Omega(\phi)$ the manifolds $W^s(p), W^u(q)$ are transversal.

Theorem 0.4.7 [Robb, Robi1, Ma2]. *A diffeomorphism ϕ is structurally stable if and only if ϕ satisfies Axiom A and the geometric STC.*

Remark. J.Robbin proved in [Robb] that the conditions are sufficient for the structural stability for diffeomorphism of class C^2, C.Robinson did the same for diffeomorphism of class C^1. R.Mañé proved the necessity.

There is another form of the strong transversality condition: the analytic *strong transversality condition* (we call it simply the STC below in this book). Fix a diffeomorphism ϕ and define for a point $x \in M$ two linear subspaces of T_xM:

$$B^+(x) = \{v \in T_xM : \lim_{k\to\infty} |D\phi^k(x)v| = 0\},$$
$$B^-(x) = \{v \in T_xM : \lim_{k\to-\infty} |D\phi^k(x)v| = 0\}.$$

Let I be a compact ϕ-invariant subset of M. We say that ϕ satisfies the STC on I if for any $x \in I$ we have

$$B^+(x) + B^-(x) = T_xM.$$

If $I = M$ we say simply that ϕ satisfies the STC.

R.Mañé proved the following result in [Ma1].

Theorem 0.4.8. *A diffeomorphism ϕ satisfies Axiom A and the geometric* STC *if and only if ϕ satisfies the* STC.

It follows from this theorem and from Theorem 0.4.7 that the set of diffeomorphisms which satisfy the STC coincides with the set of structurally stable diffeomorphisms. It will be shown in this book that diffeomorphisms which satisfy the STC play an important role also in the theory of C^0-small perturbations of dynamical systems.

Below we formulate some technical results we need. For two small smooth discs D_1, D_2 and for points $y \in D_1$, $z \in D_2$ we define the angle

$$\angle(T_y D_1, T_z D_2)$$

as follows. Fix a finite open covering $\{U_i\}$, $i = 1, \ldots, k$, of M, and let $\alpha_i : U_i \to \mathbf{R}^n$ be the corresponding coordinate maps. Find $r > 0$ such that for any $x \in M$ there is $i \in \{1, \ldots, k\}$ with the following property:

$$\overline{N_r(x)} \subset U_i. \tag{0.14}$$

Find a positive $\delta = \delta(r)$ such that for any two discs $D_1 = D(\delta, x_1)$, $D_2 = D(\delta, x_2)$ with $d(x_1, x_2) < \delta$ we have $D_1, D_2 \subset N_r(x_1)$.

Now take two discs $D_1(\Delta, x)$, $D_2(\Delta, x')$ with $\Delta \in (0, \delta]$ and with $d(x, x') < \delta$. Fix $i \in \{1, \ldots, k\}$ such that (0.14) holds and let

$$D^0_{i,j} = \alpha_i(D_j), j = 1, 2.$$

Consider two points $y \in D_1$, $z \in D_2$ and let

$$\tilde{y} = \alpha_i(y), \tilde{z} = \alpha_i(z).$$

Define

$$\angle(T_y D_1, T_z D_2) = \min_i \angle(T_{\tilde{y}} D^0_{i,1}, T_{\tilde{z}} D^0_{i,2}),$$

here $\angle$ in the right hand side is the usual angle between two linear subspaces of $\mathbf{R}^n$, and minimum is taken over all indices $i \in \{1, \ldots, k\}$ such that (0.14) holds.

Lemma 0.4.1 [Pl5] *Assume that a diffeomorphism ϕ satisfies the STC. Then there exist positive numbers α_1, α_2 such that given $\epsilon > 0$ there is $\delta > 0$ having the property : if $p, q \in \Omega(\phi)$, $x \in W^u(q)$, $y \in W^s(p)$, and $d(x, y) < \delta$ then there are smooth discs*

$$D(\alpha_1, x) \subset W^u(q), D(\alpha_1, y) \subset W^s(p)$$

having a point z of transversal intersection such that $d(x, z) < \epsilon$ and

$$\angle(T_z D(\alpha_1, x), T_z D(\alpha_1, y)) \geq \alpha_2.$$

The following result belongs to "folklore". To prove it the reader can repeat the proof of its analogue for Morse-Smale diffeomorphisms (see Lemma 0.4.4) in [Sm1] using Lemma 0.4.1.

Lemma 0.4.2. *Let Ω_i, Ω_j be two distinct basic sets for a diffeomorphism ϕ which satisfies the STC. Then the following statements are equivalent:*

(a) $\Omega_i \to \Omega_j$;
(b) $\overline{W^u(\Omega_i)} \supset W^u(\Omega_j)$;
(c) $\overline{W^u(\Omega_i)} \cap W^u(\Omega_j) \neq \emptyset$;
(d) $\overline{W^s(\Omega_j)} \supset W^s(\Omega_i)$;
(e) $\overline{W^s(\Omega_j)} \cap W^s(\Omega_i) \neq \emptyset$.

We study in this book also a special subset of the set of diffeomorphisms which satisfy the STC - the set of Morse-Smale diffeomorphisms.

We say that a diffeomorphism ϕ is a *Morse-Smale diffeomorphism* if

(a) the nonwandering set $\Omega(\phi)$ coincides with a finite set of hyperbolic periodic points;

(b) the geometric STC holds.

It is easy to understand that $\phi \in \mathrm{Diff}^1(M)$ is a Morse-Smale diffeomorphism if and only if ϕ satisfies the STC and the set $\Omega(\phi)$ is finite (see [Pa1]).

In the case $M = S^1$ the set of Morse-Smale diffeomorphisms coincides with the set of diffeomorphisms which satisfy the STC. In this case we can characterize Morse-Smale diffeomorphisms as follows: a diffeomorphism ϕ of S^1 is a Morse-Smale diffeomorphism if and only if:

(a) $\Omega(\phi)$ is a finite set which coincides with $Per(\phi)$;

(b) periodic points of ϕ are hyperbolic (note that for a hyperbolic periodic point p of a diffeomorphism $\phi \in \mathrm{Diff}^1(S^1)$ the stable manifold $W^s(p)$ is either p or an arc of S^1, hence the geometric STC holds trivially).

It was shown by M.Peixoto [Pe] that for $\phi \in \mathrm{Diff}(S^1)$ the previous conditions (a),(b) are equivalent to the following ones:

(a′) $\mathrm{Per}(\phi) \neq \emptyset$;

(b′) any periodic point of ϕ is hyperbolic.

Now consider a Morse-Smale diffeomorphism ϕ such that

$$Per(\phi) = Fix(\phi). \tag{0.15}$$

Note that for any Morse-Smale diffeomorphism ψ and for any natural m ψ^m is a Morse-Smale diffeomorphism, and if m is large enough then for $\phi = \psi^m$ (0.15) holds.

In the case (0.15) basic sets of ϕ coincide with fixed points. Following our previous notation for two different fixed points p, q of ϕ we write $p \to q$ if $W^u(p) \cap W^s(q) \neq \emptyset$. For a fixed point p consider two sets:

$\omega(W^u(p)) = \{y = \lim_{k\to\infty} \phi^{t_k}(x_k) : x_k \in W^u(p), \lim_{k\to\infty} t_k = \infty, p$ is not a limit point of $x_k\}$; $\alpha(W^s(p)) = \{y = \lim_{k\to\infty} \phi^{t_k}(x_k) : x_k \in W^s(p), \lim_{k\to\infty} t_k = -\infty, p$ is not a limit point of $x_k\}$.

The following two statements are basic results of the so called "geometric theory of Morse-Smale systems" developed by S.Smale in [Sm1].

Lemma 0.4.3. *Assume that ϕ is a Morse-Smale diffeomorphism such that (0.15) holds. Then*

(a) for fixed points p, q, r $\quad p \to q, q \to r$ imply $p \to r$;
(b) if for fixed points $p_1, \dots, p_k$ we have
$p_1 \to p_2 \to \dots \to p_k$ then these points are distinct.

Lemma 0.4.4. *For two fixed points p, q of a Morse-Smale diffeomorphism which satisfies (0.15) the following statements are equivalent:*
(a) $p \to q$;
(b) $W^u(q) \subset \omega(W^u(p))$;
(c) $W^u(q) \cap \omega(W^u(p)) \neq \emptyset$.

An analogous result is true for $\alpha(W^s(p))$.

1. Generic Properties of Dynamical Systems

1.1 Tolerance Stability

Consider a set D of dynamical systems on M.

Definition 1.1 *We say that D is a "Γ-set" if D is considered with a topology which is not coarser than the topology on D induced by the topology of $Z(M)$ (generated by ρ_0).*

Most important examples of Γ-sets are the spaces $\mathrm{Diff}^r(M)$ of C^r-diffeomorphisms of M with the C^r-topology ($r \geq 1$).

Fix a system $\phi \in Z(M)$ and a point $x \in M$. The set $\overline{O(x, \phi)}$ is an element of M^*. We define the map $\Theta : Z(M) \to M^{**}$ by

$$\Theta(\phi) = \overline{\{\overline{O(x, \phi)} : x \in M\}}. \tag{1.1}$$

Let D be a Γ-set.

Definition 1.2 *We say that a system $\phi \in Z(M)$ is "tolerance-D-stable" if ϕ is a continuity point of the restriction $\Theta \mid_D$.*

F.Takens formulated in [Ta1] the following conjecture (he called it the tolerance stability conjecture of Zeeman).

Tolerance Stability Conjecture : *for any Γ-set D there exists a residual subset D_0 of D such that any $\phi \in D_0$ is tolerance-D-stable* .

W.White constructed in [WWh] a counterexample to this conjecture. We describe below this counterexample. We begin by constructing a set of flows - often examples for flows are more "visible" than for discrete dynamical systems.

Suppose that M is two-dimensional, and identify a coordinate neighborhood of M with $\mathbf{R}^2$. The set of flows we construct will have support in $\mathbf{R}^2$, so we forget the rest of M.

Set

$$B_0 = \{(x, y) : 0 \leq x \leq 5, 0 \leq y \leq 1\},$$

$$B_1 = \{(x, y) : 2 \leq x \leq 3, 0 \leq y \leq 1\},$$

and

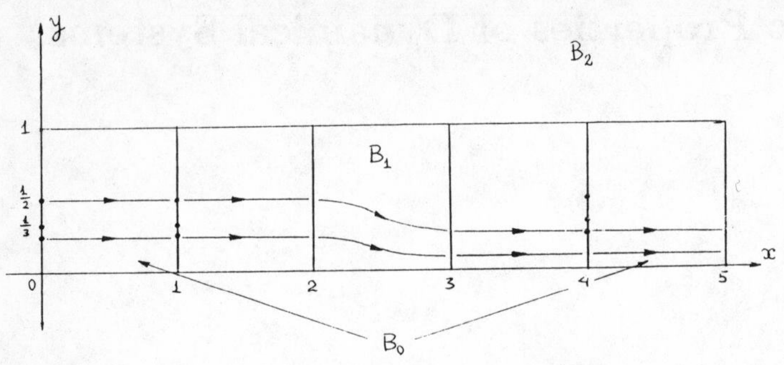

Figure 1.1

$$B_2 = (\mathbf{R}^2 \setminus \text{Int}B_0) \cup \{(i, \frac{1}{n}) : i = 1 \text{ or } 4, n = 1, 2, \ldots\}.$$

Let E be the constant vector field $p \mapsto (0,1)$ on $\mathbf{R}^2$, and let d be the usual metric on $\mathbf{R}^2$. Define D to be the set of flows Φ generated by vector fields X on $\mathbf{R}^2$ which satisfy the following conditions:

(a) $X_p = d(p, B_2)E_p$ for $p \notin \text{Int}B_1$;

(b) there is a homeomorphism $h : B_1 \to B_1$ such that h is the identity on $[2,3] \times \{0,1\}$, and $h([2,3] \times r)$ is an integral curve of $X|_{B_1}$ for each $r \in (0,1)$ (see Fig. 1.1).

We consider D with the standard C^0-topology for flows (see Sect. 1.1). Let D_1 be the set of flows in D having no orbits connecting the singular points $(1, 1/n)$ and $(4, 1/m), n, m > 0$, and let $D_2 = D \setminus D_1$. It is easy to see that both D_1, D_2 are dense in D : with a small perturbation one can break all connections between points $(1, 1/n), (4, 1/m)$ or make some connection (for large m, n).

Consider the projection $\Pi : \mathbf{R}^2 \to \mathbf{R}$ defined by $\Pi(x,y) = x$. Fix a flow $\Phi \in D_1$. It is geometrically evident that if $O(\Phi)$ is a trajectory of Φ not being a rest point, then $\Pi(\overline{O(\Phi)})$ is one of the following segments:

$$[0,1], [0,4], [0,5], [1,5].$$

Any flow $\Psi \in D_2$ has a trajectory $O(\Psi)$ such that $\Pi(\overline{O(\Psi)}) = [1,4]$. It is evident that for this trajectory $O(\Psi)$ and for any trajectory $O(\Phi), \Phi \in D_1$, we have

$$R(\overline{O(\Psi)}, \overline{O(\Phi)}) \geq 1. \tag{1.2}$$

Define the map

$$\tilde{\Theta} : D \to M^{**}$$

by

$$\tilde{\Theta}(\Phi) = \overline{\{\overline{O(\Phi)}\}}.$$

It follows from (1.2) that for any $\Phi \in D_1, \Psi \in D_2$

$$\overline{R}(\tilde{\Theta}(\Phi), \tilde{\Theta}(\Psi)) \geq 1.$$

As D_1, D_2 are dense in D, the set of continuity points of $\tilde{\Theta}$ is empty.

To obtain a corresponding set of discrete dynamical systems (we also denote this set D) take any flow $\Phi(t,p) \in D$ and let $\phi(p) = \Phi(1,p)$. It is evident that the restriction $\Theta|_D$ of the map Θ defined by (1.1) is nowhere continuous.

The Tolerance Stability Conjecture in its general form is not true. Nevertheless some results connected with this conjecture are interesting and important. We begin with results obtained by F.Takens in [Ta1]. Fix $\epsilon > 0$.

Definition 1.3 *Two systems $\phi, \psi \in Z(M)$ are "orbit-ϵ-equivalent", if:*
(1) for any $O(\phi)$ there exists $O(\psi)$ such that
(1.a) $O(\phi) \subset N_\epsilon(O(\psi))$,
(1.b) $O(\psi) \subset N_\epsilon(O(\phi))$,
(2) for any $O(\psi)$ there exists $O(\phi)$ such that
(2.a) $O(\psi) \subset N_\epsilon(O(\phi))$,
(2.b) $O(\phi) \subset N_\epsilon(O(\psi))$.

It is easy to see that if ϕ, ψ are orbit-ϵ -equivalent, then

$$\overline{R}(\Theta(\phi), \Theta(\psi)) \leq \epsilon,$$

and, conversely, if

$$\overline{R}(\Theta(\phi), \Theta(\psi)) < \epsilon,$$

then ϕ, ψ are orbit-ϵ-equivalent.

F.Takens studied two types of ϵ-equivalence.

Definition 1.4 *("min-ϵ-equivalence"). This definition is obtained from the definition of orbit-ϵ-equivalence by omitting (1.a),(2.a).*

Definition 1.5 *("max-ϵ-equivalence"). This definition is obtained from the definition of orbit-ϵ-equivalence by omitting (1.b),(2.b).*

Theorem 1.1.1 [Ta1]. *Let D be a Γ-set. There exists a residual subset $D^{\max}$ of D such that any system $\phi \in D^{max}$ has the following property: given $\epsilon > 0$ there is a neighborhood W of ϕ in D such that any two systems $\phi_1, \phi_2 \in W$ are max-ϵ-equivalent.*

Proof. Fix $\epsilon > 0$. Consider the following set: $Q_\epsilon = \{\phi \in D :$ there exists a neighborhood W of ϕ in D such that any two systems $\phi_1, \phi_2 \in W$ are max-ϵ-equivalent $\}$.

It is evident that Q_ϵ is open in D and that if

$$D^{\max} = \bigcap_{\epsilon>0} Q_\epsilon$$

then any $\phi \in D^{\max}$ has the property described in the statement of the theorem.

So to complete the proof we are to show that any Q_ϵ is dense in D. Consider an open covering $\{U_i\}, i = 1, \ldots, k$, of M such that $\operatorname{diam} U_i < \epsilon, i = 1, \ldots, k$. Let $K = \{1, \ldots, k\}$. Following our notation we denote by K^* the set of all subsets of K, and by K^{**} the set of all subsets of K^*.

Define the map

$$\Psi^{\max} : D \to K^{**}$$

by: a subset $L \subset K$ is an element of $\Psi^{\max}(\phi)$ if and only if there is an orbit $O(\phi)$ such that $O(\phi) \cap U_i \neq \emptyset$ for all $i \in L$. We first show that $\Psi^{\max}$ is lower semi-continuous. Take $\phi \in D$ and $L \in \Psi^{\max}(\phi)$. Let $L = \{l_1, \ldots, l_\nu\}$. There is a point $x \in M$ and integers $n_1, \ldots, n_\nu$ such that

$$\phi^{n_i}(x) \in U_{l_i}, i = 1, \ldots, \nu.$$

Because U_i are open there exists a neighborhood $W_L(\phi)$ of ϕ in D such that for $\tilde{\phi} \in W_L(\phi)$

$$\tilde{\phi}^{n_i}(x) \in U_{l_i}, i = 1, \ldots, \nu$$

(here we take into account that the topology on D is not coarser than the topology of $Z(M)$). Hence for $\tilde{\phi} \in W_L(\phi)$ we have $L \in \Psi^{\max}(\tilde{\phi})$. The set K^* is finite, this implies that $\Psi^{\max}$ is lower semi-continuous. It follows now from Lemma 0.2.1 that there is an open dense subset in D on which $\Psi^{\max}$ is locally constant.

Finally we show that if $\Psi^{\max}(\phi_1) = \Psi^{\max}(\phi_2)$ then ϕ_1, ϕ_2 are max-ϵ-equivalent. Take a trajectory $O(\phi_1)$ and find $L \in K^*$ such that

$$O(\phi_1) \cap U_i \neq \emptyset \text{ for } i \in L,$$

$$O(\phi_1) \subset \bigcup_{i \in L} U_i.$$

As $L \in \Psi^{\max}(\phi_1)$ we have $L \in \Psi^{max}(\phi_2)$, hence there exists a trajectory $O(\phi_2)$ such that

$$O(\phi_2) \cap U_i \neq \emptyset, i \in L.$$

Evidently

$$O(\phi_1) \subset N_\epsilon(O(\phi_2)).$$

So, ϕ_1 and ϕ_2 satisfy (1.a),(2.a) in the definition of orbit-ϵ-equivalence. This completes the proof.

An analogous result is true for min-ϵ-equivalence.

Theorem 1.1.2 [Ta1]. *Let D be a Γ-set. There exists a residual subset $D^{\min}$ of D such that any system $\phi \in D^{\min}$ has the following property : given $\epsilon > 0$ there exists a neighborhood W of ϕ in D such that any two systems $\phi_1, \phi_2 \in W$ are min-ϵ-equivalent.*

To prove this theorem fix again $\epsilon > 0$ and consider a closed covering $\{U_i\}, i \in K = \{1, \ldots, k\}$ of M with $\operatorname{diam} U_i < \epsilon$. Define

$$\Psi^{\min} : D \to K^{**}$$

by : a subset $L \subset K$ is an element of $\Psi^{\min}(\phi)$ if and only if there is an orbit $O(\phi)$ such that

$$O(\phi) \subset \bigcup_{i \in L} U_i.$$

We claim that the map $\Psi^{\min}$ is upper semi-continuous. Take $\phi \in D$ and $L \notin \Psi^{\min}(\phi)$. For any $x \in M$ there is $i(x)$ with

$$\phi^{i(x)} \notin \bigcup_{i \in L} U_i.$$

The sets $M, \bigcup_{i \in L} U_i$ are compact, hence there exist numbers $T > 0$ and $\delta > 0$ such that for any $x \in M$ there is $i(x)$ with $| i(x) | \leq T$, and

$$d(\phi^{i(x)}(x), \bigcup_{i \in L} U_i) > \delta.$$

There is a neighborhood $W_L(\phi)$ such that for any $\tilde{\phi} \in W_L(\phi)$ the same inequalities hold with $\delta/2$ instead of δ. Hence, for $\tilde{\phi} \in W_L(\phi)$

$$L \notin \Psi^{\min}(\tilde{\phi}),$$

so $\Psi^{\min}$ is upper semi-continuous. The rest of the proof is as in the case of max-ϵ-equivalence.

Remark. One can consider relations between dynamical systems analogous to orbit-ϵ-equivalence etc taking positive semi-trajectories $O^+(\phi)$ instead of trajectories $O(\phi)$. We shall use later statements for positive semi-trajectories parallel to Theorems 1.1.1, 1.1.2. Let us formulate one of them.

Let D be a Γ-set. There exists a residual subset $D_+^{\max}$ of D such that any system $\phi \in D_+^{\max}$ has the following property : given $\epsilon > 0$ there is a neighborhood W of ϕ in D such that for any $\phi_1, \phi_2 \in W$ and for any $x \in M$ we can find $y \in N_\epsilon(x)$ with

$$O^+(x, \phi_1) \subset N_\epsilon(O^+(y, \phi_2)).$$

To prove this statement fix $\epsilon > 0$ and consider an open covering $\{U_i\}$, $i \in K$, described in the proof of Theorem 1.1.1. Define maps

$$\Psi_i^{\max} : D \to K^{**}, i \in K,$$

as follows. A subset $L = (l_1, \ldots, l_m)$ is an element of $\Psi_i^{\max}(\phi)$, $\phi \in D$, if and only if there is $x \in U_i$ and numbers $t_1, \ldots, t_m > 0$ such that $\phi^{t_j}(x) \in U_{l_j}$. The same reasons as in the proof of Theorem 1.1.1 show that any $\Psi_i^{\max}$ is lower semi-continuous. Hence there is an open dense subset of D on which any $\Psi_i^{\max}$ is locally constant. To complete the proof one can repeat the end of the proof of Theorem 1.1.1.

The following result connected with the Tolerance Stability Conjecture was obtained recently by K.Odani [O] and by the author [Pi6] (note that the corresponding statement in [Pi6] is to be corrected). We remind that CLD(M) is the closure of $\text{Diff}^1(M)$ in $Z(M)$.

Theorem 1.1.3 [O, Pi6]. *A generic dynamical system ϕ in CLD(M) is tolerance-$Z(M)$-stable.*

As CLD(M)= $Z(M)$ if dim$M \leq 3$ the following theorem is a corollary of Theorem 1.1.3.

Theorem 1.1.4. *If dimM ≤ 3 then a generic dynamical system in $Z(M)$ is tolerance-$Z(M)$-stable.*

For $M = S^1$ Theorem 1.1.3 was proved earlier by K.Yano [Y1]. We give a proof of Theorem 1.1.3 later in Sect. 2.4.

1.2 Pseudotrajectories

Fix a system $\phi \in Z(M)$ and a number $\delta > 0$.

Definition 1.6 *We say that a countable set of points of M*

$$\xi = \{x_k : k \in \mathbf{Z}\}$$

is a "δ-trajectory" of ϕ if for $k \in \mathbf{Z}$

$$d(x_{k+1}, \phi(x_k)) < \delta. \tag{1.3}$$

Sometimes a δ-trajectory is called a *pseudotrajectory* (a δ-pseudotrajectory) or a pseudoorbit (a δ-pseudoorbit). We consider below also δ-semi-trajectories and finite δ-trajectories (sometimes called δ-chains).

Consider a set $\xi = \{x_k : k \geq 0\}$. We say that ξ is a δ-semi-trajectory if (1.3) holds for all $k \geq 0$.

Consider a set $\xi = \{x_k : k_1 \leq k \leq k_2\}$. We say that ξ is a finite δ-trajectory if (1.3) holds for $k_1 \leq k < k_2$.

For $x, y \in M$ we say that there is a δ-trajectory from x to y if there exists a δ-trajectory $\xi = \{x_k\}$ and $m > 0$ such that $x_0 = x, x_m = y$.

We denote for $x \in M, \phi \in Z(M)$ by CH(x, ϕ) the set of all points $y \in M$ having the following property : for any $\delta > 0$ there is a δ-trajectory from x to y.

Pseudotrajectories play very important role in the qualitative theory of dynamical systems (see [Con, Cor, Robi2]). We discuss some connections between pseudotrajectories and numerical methods for dynamical systems in Appendix C of this book.

Let us introduce the following property of a system $\phi \in Z(M)$.

Definition 1.7 *We say that $\phi \in Z(M)$ has the POTP (the "pseudoorbit tracing property") if for any $\epsilon > 0$ there is $\delta > 0$ such that if $\xi = \{x_k\}$ is a δ-trajectory of ϕ then there exists $x \in M$ with*

$$d(x_k, \phi^k(x)) < \epsilon, k \in \mathbf{Z}. \tag{1.4}$$

If (1.4) holds it is usually said that the δ-trajectory ξ is ϵ-traced by the trajectory $O(x, \phi)$. Sometimes it is said in this case that the δ- trajectory ξ is ϵ-shadowed by $O(x, \phi)$ (and the corresponding property of ϕ is called the *shadowing property*).

It was first noticed independently by D.Anosov [An2] and by R.Bowen [Bow] that if I is a hyperbolic set of a diffeomorphism ϕ then in a neighborhood of I the system ϕ has the shadowing property (we prove a version of this result in Lemma A.2, see Appendix A).

K.Odani established the following result.

Theorem 1.2.1 [O]. *A generic system $\phi \in$CLD(M) has the POTP.*

Corollary *If $dimM \leq 3$ then a generic system $\phi \in Z(M)$ has the POTP.*

Note that the genericity of the POTP for $M = S^1$ was proved by K.Yano in [Y2]. We give a proof of Theorem 1.2.1 in the end of Appendix A.

We consider also the following property POTP$^+$ being an analogue of the POTP for positive semi-trajectories.

Definition 1.8 *We say that $\phi \in Z(M)$ has the POTP$^+$ if for any $\epsilon > 0$ there is $\delta > 0$ such that if $\xi = \{x_k : k \geq 0\}$ is a δ- semi-trajectory of ϕ then there exists $x \in M$ with*

$$d(x_k, \phi^k(x)) < \epsilon, k \geq 0.$$

The following statement we use below in this book is an analogue of Theorem 1.2.1.

Theorem 1.2.1$'$. *A generic system $\phi \in$ CLD(M) has the POTP$^+$.*

One can prove this result using the same reasons we use to prove Theorem 1.2.1.

Take a δ-trajectory $\{x_k\}$ of ϕ, and take $\epsilon > 0$.

Definition 1.9 *We say that $\{x_k\}$ is "weakly ϵ-traced" by a trajectory $O(\phi)$ if*

$$\{x_k\} \subset N_\epsilon(O(\phi)). \tag{1.5}$$

Theorem 1.2.2 [Cor]. *A generic system $\phi \in Z(M)$ has the following property : given $\epsilon > 0$ there is $\delta > 0$ such that any δ- trajectory of ϕ is weakly ϵ-traced by some trajectory of ϕ.*

Remark. [1] Of course, if a system ϕ has the POTP then it has the property described in Theorem 1.2.2, so if $\dim M \leq 3$ then Theorem 1.2.2 is a corollary of Theorem 1.2.1.

Remark. [2] Let us give an example of a system which does not have the property described in Theorem 1.2.2. Take $M = S^1$ (with coordinate $\alpha \in [0,1)$), and consider $\phi \in Z(M)$ generated by $\phi(\alpha) = \alpha$. Fix $\delta > 0$, and take $\psi \in Z(M)$:

$$\psi(\alpha) = \alpha + \frac{\delta}{2} \pmod 1.$$

Evidently trajectories of ψ are δ- trajectories of ϕ, and for any $\delta \in (0, 1/2)$, $x, y \in M$ we have

$$O(y,\psi) \not\subset N_{1/4}(O(x,\phi)).$$

To prove Theorem 1.2.2 we begin by proving two lemmas (using ideas close to techniques of [Wa2]).

Lemma 1.2.1. *Let $\phi \in Z(M)$. Let $m \geq 0$ be an integer, and let $\delta > 0$ and $\eta > 0$ be given. Then for any set of points $\{x_0, \ldots, x_m\}$ such that (1.3) holds for $0 \leq k \leq m-1$ there exists a set of points $\{\xi_0, \ldots, \xi_m\}$ such that*

(a) $d(x_k, \xi_k) < \eta$, $0 \leq k \leq m$;
(b) $d(\phi(\xi_k), \xi_{k+1}) < 2\delta$, $0 \leq k \leq m-1$;
(c) $\xi_i \neq \xi_j$ for $0 \leq i < j \leq m$.

Proof. We use induction on m. For $m = 0$ the lemma is true. Suppose that lemma is true for $m-1$ and we shall prove it for m. Let $\delta > 0$ and $\eta > 0$ be given. We can suppose $\eta < \delta$. Choose $\lambda > 0$ such that $d(x,y) < \lambda$ implies $d(\phi(x), \phi(y)) < \delta$ and $\lambda < \eta$. Let $\{x_0, \ldots, x_m\}$ be given so that (1.3) holds for $0 \leq k \leq m-1$. By assumption we can choose $\{\xi_0, \ldots, \xi_{m-1}\}$ so that $d(\xi_k, x_k) < \lambda$ $(0 \leq k \leq m-1)$; $d(\phi(\xi_k), \xi_{k+1}) < 2\delta$ $(0 \leq k \leq m-2)$, and $\xi_i \neq \xi_j$ if $0 \leq i < j \leq m-1$. We know that

$$d(\phi(\xi_{m-1}), x_m) \leq d(\phi(\xi_{m-1}), \phi(x_{m-1})) + \\ + d(\phi(x_{m-1}), x_m) < 2\delta,$$

so choose ξ_m so that $\xi_m \neq \xi_i$ for $0 \leq i \leq m-1$, $d(\xi_m, x_m) < \eta$, and $d(\phi(\xi_{m-1}), \xi_m) < 2\delta$.

Lemma 1.2.2. *Let $\dim M \geq 2$. Let $\phi \in Z(M)$. Given $\Delta > 0$ there is $\delta > 0$ such that if for a set $\{x_0, \ldots, x_m\}$ (1.3) holds for $0 \leq k \leq m-1$ then there exists $\psi \in Z(M)$ and $\xi \in M$ with*

$$\rho_0(\phi, \psi) < \Delta,$$

$$d(\psi^k(\xi), x_k) < \Delta, k = 0, \dots, m. \tag{1.6}$$

Proof. Fix $\phi \in Z(M)$ and $\Delta > 0$. Find $\delta_1 \in (0, \Delta)$ such that $d(x, y) < \delta_1$ implies $d(\phi^{-1}(x), \phi^{-1}(y)) < \Delta$. Suppose that for a set $\{x_0, \dots, x_m\}$ (1.3) holds with $\delta = \delta_1/4$. By Lemma 1.2.1 there exists a set $\{\xi_0, \dots, \xi_m\}$ such that

$$d(\phi(\xi_k), \xi_{k+1}) < \delta_1/2, k = 0, \dots, m-1;$$

$$d(\xi_k, x_k) < \Delta, k = 0, \dots, m; \tag{1.7}$$

and

$$\xi_i \neq \xi_j, 0 \leq i < j \leq m$$

(hence $\phi(\xi_i) \neq \phi(\xi_j)$, $0 \leq i < j \leq m$).

It follows from Lemma 0.3.3 that there exists a diffeomorphism f having the following properties :

$$\rho_0(f, \text{ id }) < \delta_1;$$

$$f(\phi(\xi_k)) = \xi_{k+1}, k = 0, \dots, m-1. \tag{1.8}$$

Consider now $\psi = f \circ \phi$, and take $\xi = \xi_0$. Evidently (1.7) and (1.8) imply (1.6).

For any $x \in M$

$$d(\psi(x), \phi(x)) = d(f(\phi(x)), \phi(x)) < \delta_1 < \Delta,$$

$$d(\psi^{-1}(x), \phi^{-1}(x)) = d(\phi^{-1}(f^{-1}(x)), \phi^{-1}(x)) < \Delta$$

(we take into account that

$$d(f^{-1}(x), x) \leq \rho_0(f, \text{ id }) < \delta_1).$$

This completes the proof.

Let us prove Theorem 1.2.2 now. It was mentioned earlier that if $\dim M \leq 3$ then Theorem 1.2.2 is a corollary of Theorem 1.2.1, so we may consider the case $\dim M > 3$ (and so we can apply Lemma 1.2.2). Let $Z^{\max}$ be a residual subset of $Z(M)$ having the property described in Theorem 1.1.1 (we take $Z(M)$ as D) .

Take $\phi \in Z^{\max}$ and arbitrary $\epsilon > 0$. Use Theorem 1.1.1 to find $\delta_1 > 0$ such that any system $\psi \in Z(M)$ with $\rho_0(\phi, \psi) < \delta_1$ is max-$\epsilon/4$-equivalent to ϕ. Apply Lemma 1.2.2 and find for ϕ a number $\delta \in (0, \min(\delta_1, \epsilon/4))$ corresponding to $\Delta = \min(\delta_1, \epsilon/4)$. We claim that this δ has the required property. Take a δ-trajectory $\pi = \{x_k\}$ of ϕ. Let $\{M_j\}$, $j = 1, \dots, N$, be a finite open covering of M with $\text{diam} M_j < \epsilon/4$, $j = 1, \dots, N$. Fix the subset $\nu \subset \{1, \dots, N\}$ such that

$$\pi \cap M_j \neq \emptyset \text{ for } j \in \nu; \pi \cap M_j = \emptyset \text{ for } j \notin \nu.$$

Find for any $j \in \nu$ an index $k(j)$ with $x_{k(j)} \in M_j$, and let

$$k_- = \min_{j\in\nu} k(j), k_+ = \max_{j\in\nu} k(j).$$

Consider the following finite subset of π :

$$\rho = \{x_k : k_- \le k \le k_+\},$$

evidently

$$\pi \subset \overline{N_{\epsilon/4}(\rho)}. \tag{1.9}$$

Change indices of x_k so that $k_- = 0$, let in this case $m = k_+$.

Apply Lemma 1.2.2 and find for the set $\{x_0, \ldots, x_m\}$ a system ψ and a point ξ with

$$\rho_0(\phi, \psi) < \Delta \le \delta_1,$$
$$d(\psi^k(\xi), x_k) < \Delta \le \epsilon/4, k = 0, \ldots, m. \tag{1.10}$$

It follows from (1.10) that

$$\rho \subset N_{\epsilon/4}(O(\xi, \psi)).$$

The system ψ is max-$\epsilon/4$-equivalent to ϕ , hence there exists $x \in M$ with

$$O(\xi, \psi) \subset N_{\epsilon/4}(O(x, \phi)). \tag{1.11}$$

Now (1.9) - (1.11) imply

$$\pi \subset N_\epsilon(O(x, \phi)).$$

Remark. We shall use later the following statement being an analogue of Theorem 1.2.2.

Theorem 1.2.2$'$. *A generic system $\phi \in Z(M)$ has the property : given $\epsilon > 0$ there exists $\delta > 0$ such that for any δ-trajectory $\{x_k\}$ there is a positive semi-trajectory $O^+(\phi)$ with*

$$\{x_k : k \ge 0\} \subset N_\epsilon(O^+(\phi)).$$

Its proof is similar to the proof of Theorem 1.2.2, so we don't give it here.

F.Takens introduced in [Ta2] the notion of an extended orbit for a system $\phi \in Z(M)$ based on using of pseudotrajectories. He showed that if extended orbits take place of orbits in the Tolerance Stability Conjecture then the conjecture becomes true. We prove below this result of Takens. Fix $\phi \in Z(M)$.

Definition 1.10 *We say that a compact subset A of M is an "extended orbit" of ϕ if for each $\epsilon > 0$ and $\delta > 0$ there exists a δ-trajectory A_δ of ϕ such that*

$$R(A, \overline{A_\delta}) < \epsilon.$$

Define $E(\phi) \subset M^*$ to be the set of all $A \in M^*$ being extended orbits for ϕ; it follows that $E(\phi)$ is closed and hence may be considered as a point of M^{**}. Consider the map

$$E : Z(M) \to M^{**}$$

which assigns to each ϕ the point $E(\phi)$.

Theorem 1.2.3 [Ta2]. *For any Γ-set D there exists a residual subset D_0 of D such that any $\phi \in D_0$ is a continuity point of the restriction $E\,|_D$.*

Proof. It follows from Lemma 0.2.1 that it is enough to show that the map $E : D \to M^{**}$ is upper semi-continuous. Take $\phi \in D$ and $\epsilon > 0$. For $\delta > 0$ define $E(\phi, \delta) \in M^{**}$ to be the closure of the set of those points $A \in M^*$, which are, considered as closed subsets of M, closures of δ-trajectories of ϕ. From the definition of $E(\phi)$ it follows that

$$E(\phi) = \bigcap_{\delta>0} E(\phi, \delta).$$

As M^* is compact, and $E(\phi, \delta_1) \subset E(\phi, \delta)$ for $0 < \delta_1 \leq \delta$, it follows that there is $\delta_0 > 0$ such that for every $\delta \in (0, \delta_0]$

$$E(\phi, \delta) \subset N_\epsilon(E(\phi)).$$

Now we take a neighborhood V of ϕ in D defined by

$$V = \{\psi \in D : d(\phi(x), \psi(x)) < \delta_0/2, x \in M\}.$$

The fact that V is open in D follows from the fact that the topology on D is not coarser than the topology induced by the topology of $Z(M)$. It is evident that if $\psi \in V$ and A is a closure of a δ-trajectory for ψ , then A is a closure of a $(\delta_0/2 + \delta)$-trajectory for ϕ. Hence, for $\psi \in V$ and for $\delta \in (0, \delta_0/2)$

$$E(\psi, \delta) \subset E(\phi, \delta_0) \subset N_\epsilon(E(\phi)).$$

Taking into account that

$$E(\psi) = \bigcap_{\delta>0} E(\psi, \delta)$$

we see that

$$E(\psi) \subset N_\epsilon(E(\phi))$$

for any $\psi \in V$. This completes the proof.

1.3 Prolongations

Various types of prolongations of trajectories of dynamical systems were studied by many authors [An, D1, D2]. Usually prolongations are a tool to investigate stability of trajectories.

For dynamical systems orbital stability of positive semi-trajectories is often considered instead of Lyapunov stability.

Fix a system $\phi \in Z(M)$ and a point $x \in M$.

Definition 1.11 *We say that the positive semi-trajectory $O^+(x,\phi)$ is "orbitally stable" if given $\epsilon > 0$ there exists $\delta > 0$ such that for any $y \in N_\delta(x)$*

$$O^+(y,\phi) \subset N_\epsilon(O^+(x,\phi))$$

Another type of stability is the so-called stability with respect to permanent perturbations. P.Bohl [Boh] was the first who studied this type of stability for solutions of differential equations.

Definition 1.12 *We say that $O^+(x,\phi)$ is "stable with respect to permanent perturbations" if given $\epsilon > 0$ there exists $\delta > 0$ such that for any system $\psi \in N_\delta(\phi)$*

$$O^+(x,\psi) \subset N_\epsilon(O^+(x,\phi)).$$

Let us now define main types of prolongations. Fix again $\phi \in Z(M), x \in M$, and consider the following sets:

$$Q(x,\phi) = \bigcap_{\eta>0} \overline{\bigcup_{y \in N_\eta(x)} O^+(y,\phi)};$$

$$P(x,\phi) = \bigcap_{\eta>0} \overline{\bigcup_{\psi \in N_\eta(\phi)} O^+(x,\psi)};$$

$$R(x,\phi) = \{x\} \cup \mathrm{CH}(x,\phi)$$

(the set $\mathrm{CH}(x,\phi)$ was defined in the previous section).

We call $Q(x,\phi)$ the *prolongation* of x in ϕ *with respect to the initial point*, $P(x,\phi)$ the *prolongation* of x in ϕ *with respect to the system*, and $R(x,\phi)$ the *chain prolongation* of x in ϕ.

Standard considerations show that the sets $Q(x,\phi), P(x,\phi), R(x,\phi)$ are closed, and that

$$O^+(x,\phi) \subset Q(x,\phi) \cap P(x,\phi) \cap R(x,\phi).$$

It is easy to see that $O^+(x,\phi)$ is orbitally stable if and only if

$$Q(x,\phi) = \overline{O^+(x,\phi)}, \tag{1.12}$$

and that $O^+(x,\phi)$ is stable with respect to permanent perturbations if and only if

$$P(x,\phi) = \overline{O^+(x,\phi)}. \tag{1.13}$$

For $\phi \in Z(M)$ denote by $L(\phi)$ the set of points $x \in M$ for which (1.13) is true. V.Dobrynsky and A.Sharkovsky established the following result.

Theorem 1.3.1 [D1]. *For a generic system $\phi \in Z(M)$ the set $L(\phi)$ is residual in M.*

To prove this theorem we need some intermediate lemmas.

Lemma 1.3.1. *Assume that for $\phi \in Z(M), x \in M$ and for some $\epsilon > 0$*

$$P(x,\phi) \subset N_\epsilon(O^+(x,\phi)). \tag{1.14}$$

Then there exists $\eta > 0$ such that

$$P(y,\psi) \subset N_\epsilon(O^+(y,\psi))$$

for any $y \in N_\eta(x), \psi \in N_\eta(\phi)$.

Proof. To obtain a contradiction suppose that there exist points x_k and systems ϕ_k such that

$$x_k \to x, \phi_k \to \phi \text{ as } k \to \infty,$$

and

$$P(x_k,\phi_k) \not\subset N_\epsilon(O^+(x_k,\phi_k)).$$

Take points $y_k \in P(x_k,\phi_k)$ such that

$$d(y_k, O^+(x_k,\phi_k)) \geq \epsilon, \tag{1.15}$$

and let y be a limit point of the sequence y_k. It is easy to see that for any $m \geq 0$

$$d(y, \phi^m(x)) \geq \epsilon. \tag{1.16}$$

Indeed, if for some $m \geq 0$ we have $d(y,\phi^m(x)) < \epsilon$ then for large k $d(y_k, \phi_k^m(x_k)) < \epsilon$, that contradicts to (1.15).

As $y_k \in P(x_k,\phi_k)$ there exist systems ψ_k and numbers $m_k \geq 0$ having the following properties :

$$\lim_{k\to\infty} \rho_0(\phi_k,\psi_k) = 0, y_k = \lim_{k\to\infty} \psi_k^{m_k}(x_k).$$

It follows from Lemma 0.3.2 that there exist systems $\tilde{\psi}_k$ and numbers $\mu_k \geq 0$ such that

$$\lim_{k\to\infty} \rho_0(\tilde{\psi}_k,\phi) = 0, \tilde{\psi}_k^{\mu_k}(x) = \psi_k^{m_k}(x_k).$$

Finally we see that $y_k \in P(x,\phi)$. The set $P(x,\phi)$ is closed, hence $y \in P(x,\phi)$. So we obtained a contradiction between (1.14) and (1.16).

We remind that as usually G_δ is the class of subsets of a topological space being countable intersections of open sets.

Lemma 1.3.2. *For any system $\phi \in Z(M)$ the set $L(\phi)$ is of class G_δ.*

Proof. Define the following sets L_ϵ for $\epsilon > 0 : x \in L_\epsilon$ if and only if there exists a neighborhood $U(x)$ of x such that

$$P(y,\phi) \subset N_\epsilon(O^+(y,\phi)) \tag{1.17}$$

for any $y \in U(x)$. It follows from the definition that any L_ϵ is open. We claim that

$$L(\phi) = \bigcap_{n=1}^{\infty} L_{\frac{1}{n}}.$$

If $x \notin L(\phi)$ then there exists $\epsilon > 0$ such that

$$P(x,\phi) \not\subset N_\epsilon(O^+(x,\phi)),$$

hence, $x \notin L_\epsilon$. Take now $x \in L(\phi)$, then (1.14) holds for any $\epsilon > 0$. It follows from Lemma 1.3.1 that there exists $\eta > 0$ such that (1.17) holds for $y \in N_\eta(x)$, so $x \in L_\epsilon$.

Let

$$Z_1 = \{\phi \in Z(M) : \overline{L(\phi)} = M\}.$$

Lemma 1.3.3. *The set Z_1 is of class G_δ.*

Proof. Fix a countable family of open sets $\Gamma_j, j = 1, 2, \ldots$, being a basis for the topology of M. It is evident that $\phi \notin Z_1$ if and only if there exists j such that

$$P(x,\phi) \neq \overline{O^+(x,\phi)}$$

for any $x \in \Gamma_j$.

Let Γ be an open subset of M, and take $\epsilon > 0$. Consider the set

$$B_{\Gamma,\epsilon} = \{\phi \in Z(M) : P(x,\phi) \not\subset N_\epsilon(O^+(x,\phi)), x \in \Gamma\}.$$

We claim that $B_{\Gamma,\epsilon}$ is closed in $Z(M)$. Indeed, consider a sequence $\phi_m \in B_{\Gamma,\epsilon}$, and let ϕ be a limit point of this sequence. If $\phi \notin B_{\Gamma,\epsilon}$ then there exists $x \in \Gamma$ such that (1.14) holds. Apply Lemma 1.3.1 to see that in this case

$$P(x,\phi_m) \subset N_\epsilon(O^+(x,\phi_m))$$

for large m. The contradiction with the choice of ϕ_m we obtained shows that $B_{\Gamma,\epsilon}$ is closed. Hence, the set

$$B = \bigcup_{j=1}^{n} \bigcup_{n=1}^{\infty} B_{\Gamma_j,\frac{1}{n}} \tag{1.18}$$

is a countable union of closed sets.

To complete the proof we show now that

$$B = Z(M) \setminus Z_1. \tag{1.19}$$

It follows immediately from definitions that for any j, n

$$Z_1 \cap B_{\Gamma_j, \frac{1}{n}} = \emptyset,$$

so $B \subset Z(M) \setminus Z_1$.

Let us prove that $Z(M) \setminus Z_1 \subset B$. For a system $\phi \in Z(M) \setminus Z_1$ there exists an open set U such that

$$U \subset \{x \in M : P(x, \phi) \neq \overline{O^+(x, \phi)}\}.$$

To obtain a contradiction suppose that $\phi \notin B$ then $\phi \notin B_{\Gamma_j, \frac{1}{n}}$ for any j, n. Find j_0 such that $\Gamma_{j_0} \subset U$.

As $\phi \notin B_{\Gamma_{j_0}, 1}$ we can take $x \in \Gamma_{j_0}$ with

$$P(x, \phi) \subset N_1(O^+(x, \phi)).$$

It follows from Lemma 1.3.1 that there exists $\eta_1 > 0$ such that

$$P(y, \phi) \subset N_1(O^+(y, \phi))$$

for any $y \in N_{\eta_1}(x)$. Find j_1 such that

$$\overline{\Gamma_{j_1}} \subset N_{\eta_1}(x) \cap \Gamma_{j_0}.$$

As $\phi \notin B_{\Gamma_{j_1}, \frac{1}{2}}$ there exists $x_1 \in \Gamma_{j_1}$ with

$$P(x_1, \phi) \subset N_{\frac{1}{2}}(O^+(x_1, \phi)),$$

find $\eta_2 > 0$ such that

$$P(y, \phi) \subset N_{\frac{1}{2}}(O^+(y, \phi))$$

for any $y \in N_{\eta_2}(x_1)$. Continuing this process we obtain a sequence

$$\Gamma_{j_0} \supset \overline{\Gamma_{j_1}} \supset \Gamma_{j_1} \supset \overline{\Gamma_{j_2}} \supset \dots$$

having the property :

$$P(x, \phi) \subset N_{\frac{1}{k}}(O^+(x, \phi))$$

for any $x \in \Gamma_{j_k}$. If we take now

$$x \in \bigcap_{k=1}^{\infty} \overline{\Gamma_{j_k}}$$

then evidently (1.13) holds. The contradiction with the choice of U completes the proof.

Lemma 1.3.4. *The set Z_1 is dense in $Z(M)$.*

Proof. Taking into account (1.18), (1.19) it is enough to show that any set $B_{\Gamma,\epsilon}$ is nowhere dense in M. As $B_{\Gamma,\epsilon}$ are closed we shall show that any $B_{\Gamma,\epsilon}$ contains no open subsets of M. Fix $B_{\Gamma,\epsilon}$ and let $U_1, \ldots, U_k$ be an open covering of M with $\operatorname{diam} U_i < \epsilon$, $i = 1, \ldots, k$. Take $\phi \in B_{\Gamma,\epsilon}$, and $\eta > 0$. We claim that

$$N_\eta(\phi) \setminus B_{\Gamma,\epsilon} \neq \emptyset. \tag{1.20}$$

Let x_0 be an arbitrary point of Γ. Then there exists $i_1 \in \{1, \ldots, k\}$ such that $x_0 \in U_{i_1}$. As $\phi \in B_{\Gamma,\epsilon}$, and $O^+(x_0, \phi) \cap U_{i_1} \neq \emptyset$ we see that

$$P(x_0, \phi) \not\subset U_{i_1}. \tag{1.21}$$

It follows from (1.21) that there exists $i_2 \neq i_1$ with

$$P(x_0, \phi) \cap U_{i_2} \neq \emptyset.$$

Take $\eta_1 < \frac{\eta}{k}$ and find a system $\phi_1 \in N_{\eta_1}(\phi)$ such that

$$O^+(x_0, \phi_1) \cap U_{i_2} \neq \emptyset. \tag{1.22}$$

If

$$P(x_0, \phi_1) \subset N_\epsilon(O^+(x_0, \phi_1))$$

then $\phi_1 \notin B_{\Gamma,\epsilon}$, so (1.20) is true. If $\phi_1 \in B_{\Gamma,\epsilon}$ then

$$P(x_0, \phi_1) \not\subset U_{i_1} \cup U_{i_2}$$

(take into account inclusions $x_0 \in U_{i_1}$ and (1.22)). Find $i_3 \neq i_1, i_2$ such that

$$P(x_0, \phi_1) \cap U_{i_3} \neq \emptyset.$$

There exists $\eta_2 < \frac{\eta}{k}$ such that for any system $\phi_2 \in N_{\eta_2}(\phi_1)$

$$O^+(x_0, \phi_2) \cap U_{i_2} \neq \emptyset, O^+(x_0, \phi_2) \cap U_{i_3} \neq \emptyset,$$

and so on. The described process has not more than k steps, and finally we obtain a system belonging to $N_\eta(\phi)$ (as $\eta_1 + \eta_2 + \ldots < \eta$), and not belonging to $B_{\Gamma,\epsilon}$. Hence, (1.20) is true. This completes the proof.

Theorem 1.3.1 is an immediate corollary of Lemmas 1.3.1 -1.3.4. Now we prove the following result concerning properties of prolongations in generic dynamical systems.

Theorem 1.3.2. *There exists a residual subset Z^* of $Z(M)$ such that*

$$P(x, \phi) = Q(x, \phi) = R(x, \phi) \tag{1.23}$$

for any $\phi \in Z^*, x \in M$.

Remark. V.Dobrynsky showed in [D2] that for a generic system $\phi \in Z(M)$ $P(x,\phi) = Q(x,\phi)$ for any $x \in M$, the complete result was published by the author in [Pi5].

Proof. Let us first show that for any $\phi \in Z(M), x \in M$

$$Q(x,\phi) \subset P(x,\phi). \tag{1.24}$$

Take $y \in Q(x,\phi)$. If $y = x$ then evidently $y \in P(x,\phi)$. If $y \neq x$ fix sequences $x_m \in M, k(m) \geq 0$ such that

$$\lim_{m\to\infty} x_m = x, \lim_{m\to\infty} \phi^{k(m)}(x_m) = y.$$

Apply Lemma 0.3.2 to find systems ψ_m and numbers $\lambda(m) \geq 0$ such that

$$\lim_{m\to\infty} \rho_0(\psi_m, \phi) = 0, \lim_{m\to\infty} \psi_m^{\lambda(m)}(x) = y,$$

hence $y \in P(x,\phi)$.

There exists a residual subset $Z_+^{\max}$ of $Z(M)$ having the following property : if $\phi \in Z_+^{\max}$ then given $\epsilon > 0$ one can find a neighborhood W of ϕ such that for any $x \in M$ and for any $\psi \in W$ there is $\xi \in N_\epsilon(x)$ with

$$O^+(x,\psi) \subset N_\epsilon(O^+(\xi,\phi))$$

(see the remark after the proof of Theorem 1.1.2; we apply the analogue of Theorem 1.1.1 for semi-trajectories taking $D = Z(M)$).

We claim that if $\phi \in Z_+^{\max}$ then for any $x \in M$

$$P(x,\phi) \subset Q(x,\phi). \tag{1.25}$$

Take $\phi \in Z_+^{\max}, x \in M$, and $y \in P(x,\phi)$. There exist seqences $\phi_m \in Z(M), k(m) \geq 0$ such that

$$\lim_{m\to\infty} \rho_0(\phi_m, \phi) = 0, y = \lim_{m\to\infty} \phi_m^{k(m)}(x).$$

Fix $\epsilon > 0$. There is $m_0 > 0$ such that for $m \geq m_0$ we can find $x_m \in N_\epsilon(x)$ having the following property :

$$O^+(x,\phi_m) \subset N_\epsilon(O^+(x_m,\phi)).$$

So we obtain sequences x_m, z_m with

$$z_m \in O^+(x_m,\phi), \lim_{m\to\infty} x_m = x, \lim_{m\to\infty} z_m = y,$$

hence $y \in Q(x,\phi)$.

Denote by Z_+^0 a residual set in $Z(M)$ such that systems in Z_+^0 have the property described in Theorem 1.2.2′. Let $Z^* = Z_+^0 \cap Z_+^{\max}$. Of course, Z^* is a residual subset of $Z(M)$. Take $\phi \in Z^*$ and $x \in M$. We claim that

$$R(x,\phi) \subset Q(x,\phi). \tag{1.26}$$

Take $y \in R(x,\phi)$. If $y = x$ then $y \in Q(x,\phi)$. Now consider $y \neq x$. Fix arbitrary $\epsilon > 0$. Find for this ϵ a number δ such that any δ-trajectory of ϕ is weakly ϵ-traced by a trajectory of ϕ. It follows from the definition of $R(x,\phi)$ that there is a δ-trajectory $\{x_k\}$ of ϕ such that $x_0 = x, x_m = y$ for some $m > 0$. Hence there exists a point $\xi \in N_\epsilon(x)$ and $k \geq 0$ such that

$$d(y, \phi^k(\xi)) < \epsilon.$$

So we can find sequences $x_m \in M$ and numbers $k(m) \geq 0$ with

$$\lim_{m\to\infty} x_m = x, \lim_{m\to\infty} \phi^{k(m)}(x_m) = y,$$

consequently $y \in Q(x,\phi)$. If for a dynamical system $\psi \in Z(M)$ we have $\rho_0(\phi,\psi) < \delta$ then any trajectory of ψ is a δ-trajectory of ϕ, hence

$$P(x,\phi) \subset R(x,\phi) \tag{1.27}$$

It follows from (1.24)-(1.27) that for $x \in M, \phi \in Z^*$ (1.23) holds.

It was shown in Theorem 1.3.1 that for a generic system ϕ and for a generic point $x \in M$ (1.13) is true. Applying Theorem 1.3.2 we see that for a generic system ϕ and for a generic point $x \in M$ (1.14) is also true. So, for a generic dynamical system $\phi \in Z(M)$, a generic positive semi-trajectory is both orbitally stable and stable with respect to permanent perturbations.

Now we define the so-called limit prolongations. Fix a system $\phi \in Z(M)$, a point $x \in M$ and consider the following sets :

$$Q^\omega(x,\phi) = \{y = \lim_{k\to\infty} \phi^{t_k}(x_k) : x_k \to x, t_k \to +\infty \text{ as } k \to \infty\},$$

$$P^\omega(x,\phi) = \{y = \lim_{k\to\infty} \phi_k^{t_k}(x) : \phi_k \to \phi, t_k \to +\infty \text{ as } k \to \infty\},$$

$$R^\omega(x,\phi) = \{y : \text{ there exists a sequence of } \frac{1}{k}\text{-trajectories } \{x_m^k\}$$
$$\text{of } \phi \text{ and a sequence } m_k \to +\infty \text{ such that } x_0^k = x, x_{m_k}^k = y\}.$$

It is easy to see that the sets $Q^\omega(x,\phi), P^\omega(x,\phi), R^\omega(x,\phi)$ are ϕ-invariant compact sets, and

$$S(x,\phi) = O^+(x,\phi) \cup S^\omega(x,\phi) \text{ for } S = Q, P, R. \tag{1.28}$$

Theorem 1.3.3. *For any system $\phi \in Z^*$ and for any point $x \in M$*

$$P^\omega(x,\phi) = Q^\omega(x,\phi) = R^\omega(x,\phi) \tag{1.29}$$

(here Z^* is the resudial subset of $Z(M)$ given by Theorem 1.3.2).

Proof. Considerations analogous to those we used in the proof of Theorem 1.3.2 show that for any $\phi \in Z(M), x \in M$ we have

$$Q^\omega(x,\phi) \subset P^\omega(x,\phi) \subset R^\omega(x,\phi). \tag{1.30}$$

Let us show that for $\phi \in Z^*$ and for any $x \in M$

$$R^\omega(x,\phi) \subset Q^\omega(x,\phi). \tag{1.31}$$

Take $y \in R^\omega(x,\phi)$. As $y \in R(x,\phi)$ we obtain that $y \in Q(x,\phi)$. If $y \notin O^+(x,\phi)$ then it follows from (1.28) (with $S = Q$) that $y \in Q^\omega(x,\phi)$. Now consider the case $y \in O^+(x,\phi)$, let $y = \phi^{m_0}(x)$ with some $m_0 \geq 0$.

If $x \in \mathrm{Per}(\phi)$ then $y \in \mathrm{Per}(\phi)$ and evidently $y \in Q^\omega(x,\phi)$ - take $x_k \equiv x, t_k = m_0 + kp$ where p is a positive period of x.

If $x \notin \mathrm{Per}(\phi)$ then $\phi^k(x) \neq x$ for $k \in \mathbf{Z}$, $k \neq 0$. To obtain a contradiction suppose that $y \notin Q^\omega(x,\phi)$. As the set $Q^\omega(x,\phi)$ is ϕ-invariant, and $y \in O^+(x,\phi)$ we see that $\phi^{-1}(x) \notin Q^\omega(x,\phi)$. By (1.20) $\phi^{-1}(x) \notin R^\omega(x,\phi)$. Let $\mu_k = m_k - m_0 - 1$. It follows from

$$\lim_{k\to\infty} x^k_{m_k} = y$$

that

$$\lim_{k\to\infty} x^k_{\mu_k} = \phi^{-m_0-1}(y) = \phi^{-1}(x).$$

As $x \notin \mathrm{Per}(\phi)$, $\mu_k \to \infty$, and the set

$$\{x^k_0, \dots, x^k_{\mu_k-1}, \phi^{-1}(x)\}$$

is a finite δ_k-trajectory for ϕ where

$$\delta_k = \frac{1}{k} + d(x^k_{\mu_k}, \phi^{-1}(x)),$$

$$\lim_{k\to\infty} \delta_k = 0.$$

Hence, $\phi^{-1}(x) \in Q^\omega(x,\phi) \subset R^\omega(x,\phi)$.

The contradiction we obtained proves (1.31). Now it follows from (1.30), (1.31) that for $\phi \in Z^*, x \in M$ (1.29) is true.

1.4 Returning Points and Filtrations

If we want to investigate the global structure of trajectories of a dynamical system it is very important to study trajectories "returning" to their initial points.

We mentioned earlier in Sect. 0.1 the sets $\mathrm{Fix}(\phi)$ of fixed points, $\mathrm{Per}(\phi)$ of periodic points, and the nonwandering set $\Omega(\phi)$ for a dynamical system $\phi \in Z(M)$. Let us introduce the following sets of "returning" points (in various senses) for $\phi \in Z(M)$.

The *set of weakly periodic points*

$$\mathrm{WPer}(\phi) = \{y = \lim_{k\to\infty} y_k : y_k \in Per(\phi_k), lim_{k\to\infty}\phi_k = \phi\}.$$

The *set of weakly nonwandering points*

$$\mathrm{W}\Omega(\phi) = \{y = \lim_{k\to\infty} y_k : y_k \in \Omega(\phi_k), lim_{k\to\infty}\phi_k = \phi\}.$$

The *chain-recurrent set*

$$\mathrm{CR}(\phi) = \{x : x \in CH(x, \phi)\}.$$

It is easy to see that the sets WPer(ϕ), WΩ(ϕ), CR(ϕ) are ϕ-invariant and compact, so the maps Ω, WPer, WΩ, CR:$Z(M) \to M^*$ are defined.

Definition 1.13 *We say that a system $\phi \in Z(M)$ has no C^0 Ω - explosions if given $\epsilon > 0$ there exists $\delta > 0$ such that for any $\psi \in Z(M)$ with $\rho_0(\phi, \psi) < \delta$ we have*

$$\Omega(\psi) \in N_\epsilon(\Omega(\phi)).$$

In other words, ϕ has no C^0 Ω - explosions if and only if the map Ω : $Z(M) \to M^*$ is upper semi-continuous at ϕ.

Theorem 1.4.1. *A generic system ϕ has the following properties:*

(a) ϕ has no C^0 Ω - explosions ;

(b) ϕ is a continuity point of the maps WPer, WΩ, CR;

(c) $\Omega(\phi)$ = WPer(ϕ) = WΩ(ϕ) = CR(ϕ).

Remark. F.Takens proved (a) in [Ta1]. Other statements of Theorem 1.4.1 seem to be "folklore".

Proof. Consider an open covering $\{U_i\}, i = 1, \ldots, k$, of M. For $\phi \in Z(M)$, let $\Psi(\phi)$ be the subset of $K = \{1, \ldots, k\}$ consisting of those $i \in K$ for which there is an integer $n(i) \neq 0$ with

$$\phi^{n(i)}(U_i) \cap U_i \neq 0.$$

Ψ is a map from $Z(M)$ to K^*. This map is lower semi-continuous because if $i \in \Psi(\phi)$ then for ψ close enough to ϕ, $i \in \Psi(\psi)$. Consider now an infinite sequence $U_{i,j}, j = 1, \ldots, \infty; i = 1, \ldots, k(j)$, of open coverings of M with diam$U_{i,j} < \frac{1}{j}$. Let $K_j = \{1, \ldots, k(j)\}$, and define $\Psi_j : Z(M) \to K_j^*$ as above. Each Ψ_j is lower semi-continuous, let $D \subset Z(M)$ be the set of points where all Ψ_j are continious. It follows from Lemma 0.2.1 that D is residual in $Z(M)$. We claim that elements of D have no C^0 Ω - explosions.

First we notice that for $\phi \in Z(M)$, $x \in M, x \notin \Omega(\phi)$ is equivalent with: there is a set $U_{i,j}$ such that $x \in U_{i,j}$, and $i \notin \Psi_j(\phi)$.

Now we fix $\phi \in D$ and $\epsilon > 0$. Denote by A the complement of $N_\epsilon(\Omega(\phi))$. As A is compact, it follows that there is a finite covering

$$U_{i_1,j_1},\ldots,U_{i_r,j_r}$$

of A such that for all $\nu = 1,\ldots,r, i_\nu \notin \Psi_{j_\nu}(\phi)$.

Because all maps Ψ_j are continuous at ϕ there is $\delta > 0$ such that for any $\psi \in Z(M)$ with $\rho_0(\phi,\psi) < \delta$ and for $\nu = 1,\ldots,r$ we have $i_\nu \notin \Psi_{j_\nu}(\phi)$. Hence

$$\Omega(\psi) \subset M \setminus A = N_\epsilon(\Omega(\phi)).$$

This proves the statement (a).

We begin proving the statement (b) by showing that the map WPer $:Z(M) \to M^*$ is upper semi-continuous. To obtain a contradiction suppose that there exists $\epsilon > 0$ and sequences of systems $\phi_m, \lim_{m\to\infty}\phi_m = \phi$, and points $y_m \in WPer(\phi_m)$ such that $y_m \notin N_\epsilon(WPer(\phi))$. Let y be a limit point of the sequence y_m. Then

$$d(y, WPer(\phi)) \geq \epsilon. \tag{1.32}$$

Find for any y_m a system ψ_m and a point $z_m \in Per(\psi_m)$ such that

$$\rho_0(\phi_m,\psi_m) < \frac{1}{m},$$

$$d(z_m,y_m) < \frac{1}{m}.$$

Then $\lim_{m\to\infty}\psi_m = \phi$, $\lim_{m\to\infty} z_m = y$, hence $y \in WPer(\phi)$, which contradicts to (1.32). So, the map $WPer$ is upper semi-continuous, hence, the set of its continuity points is residual in $Z(M)$.

Similar proof shows that the set of continuity points of the map $W\Omega$ is residual in $Z(M)$.

Let us now show that the map $CR : Z(M) \to M^*$ is upper semi-continuous. Take arbitrary sequences $\phi_m, \lim_{m\to\infty}\phi_m = \phi$, $y_m \in CR(\phi_m)$, and let y be a limit point of y_m. We claim that $y \in CR(\phi)$.

Take arbitrary $\delta > 0$. Find $\Delta > 0$ such that $d(x,y) < \Delta$ implies $d(\phi(x),\phi(y)) < \delta/3$. Find m such that

$$\rho_0(\phi,\phi_m) < \delta/3,$$

$$d(y_m,y) < \min(\Delta,\delta/3).$$

There exists a $\delta/3$ - trajectory $\{x_k\}$ of ϕ_m with $x_0 = y_m, x_r = y_m$ for some $r \neq 0$ (we consider the case $r > 0$). Define the following set $\{\xi_k : k \in \mathbf{Z}\}$: $\xi_0 = \xi_r = y$; $\xi_k = x_k$ for $k = 1,\ldots,r-1$; $\xi_k = \phi^k(y)$ for $k = -1,-2,\ldots$; $\xi_k = \phi^{k-r}(y)$ for $k = r+1, r+2,\ldots$. We claim that $\{\xi_k\}$ is a δ - trajectory of ϕ. If $k \notin \{0,\ldots,r-1\}$ then $d(\phi(\xi_k),\xi_{k+1}) = 0$.

Take $k = 0$. We have

$$d(\phi(\xi_0),\xi_1) \leq d(\phi(y),\phi(y_m)) + d(\phi(y_m),\phi_m(y_m)) +$$

$$+ d(\phi_m(y_m),x_1) < \delta.$$

We take into account here that $\xi_0 = y$, $d(y, y_m) < \Delta$, $\rho_0(\phi, \phi_m) < \delta/3$, $\phi_m(y_m) = \phi_m(x_0)$. Similary, for $k = 1, \dots, r-2$,

$$d(\phi(\xi_k), \xi_{k+1}) \le d(\phi_m(\xi_k), \phi(\xi_k)) + d(\phi_m(\xi_k), \xi_{k+1}) < \delta;$$

$$d(\phi(\xi_{r-1}), \xi_r) \le d(\phi(\xi_{r-1}), \phi_m(\xi_{r-1})) +$$
$$+ d(\phi_m(\xi_{r-1}), y_m) + d(y_m, y) < \delta.$$

So, $y \in CR(\phi)$. Hence, the set of continuity points of $CR : Z(M) \to M^*$ is residual in $Z(M)$. This completes the proof of (b).

To prove (c) note first that it is evident that for any $\phi \in Z(M)$

$$WPer(\phi) \subset W\Omega(\phi), \Omega(\phi) \subset W\Omega(\phi). \tag{1.33}$$

Let us show that for any $\phi \in Z(M)$

$$W\Omega(\phi) \subset CR(\phi). \tag{1.34}$$

Take $y \in W\Omega(\phi)$ and arbitrary $\delta > 0$. Find $\Delta > 0$ such that $d(x, y) < \Delta$ implies $d(\phi(x), \phi(y)) < \delta/2$. It follows from the definition of the set $W\Omega(\phi)$ that there exists a system ψ, a point ξ, and a number $m > 0$ such that

$$\rho_0(\phi, \psi) < \frac{\delta}{2};$$

$$d(\xi, y), d(\psi^m(\xi), y) < \min(\Delta, \delta/2).$$

Consider the sequence of points $\{x_k : k \in \mathbf{Z}\}$ such that $x_k = \psi^k(\xi)$ for $k \ne 0, m$; $x_0 = x_m = y$. We claim that $\{x_k\}$ is a δ - trajectory of ϕ, then it follows that $y \in CR(\phi)$.

Evidently for $k \ne -1, 0, m-1, m$ we have

$$d(\phi(x_k), x_{k+1}) = d(\phi(x_k), \psi(x_k)) < \delta/2.$$

Take $k = -1$:

$$d(\phi(x_{-1}), x_0) \le d(\phi(x_{-1}), \psi(x_{-1})) + d(\xi, y) < \delta.$$

Take now $k = 0$:

$$d(\phi(x_0), x_1) \le d(\phi(y), \phi(\xi)) + d(\phi(\xi), \psi(\xi)) < \delta$$

(we take into account that $d(y, \xi) < \Delta$ here). Similary,

$$d(\phi(x_k), x_{k+1}) < \delta \text{ for } k = m-1, m.$$

So, we showed that (1.34) is true.

Let Z^* be a residual subset of $Z(M)$ such that for $\phi \in Z^*$ the statement of Theorem 1.3.2 is true.

We claim that for $\phi \in Z^*$

$$CR(\phi) \subset \Omega(\phi). \tag{1.35}$$

Take $y \in CR(\phi)$. If $\phi(y) = y$, then $y \in \Omega(\phi)$. Consider now $y \in CR(\phi)$ such that $y_1 = \phi(y) \neq y$. It follows evidently that $y \in R(y_1, \phi)$. As $\phi \in Z^*$ we have $y \in Q(y_1, \phi)$. Hence there exist sequences $\xi_m, \lim_{m\to\infty} \xi_m = y_1$, $k(m) \geq 0$ such that

$$\lim_{m\to\infty} \phi^{k(m)}(\xi_m) = y.$$

Let $z_m = \phi^{-1}(\xi_m)$. Then $\lim_{m\to\infty} z_m = y$ and

$$\lim_{m\to\infty} \phi^{k(m)+1}(z_m) = y,$$

hence $y \in \Omega(\phi)$.

To complete the proof let us show that for any $\phi \in Z(M)$

$$\Omega(\phi) \subset WPer(\phi). \tag{1.36}$$

Consider $y \in \Omega(\phi)$, and find sequences $\xi_m, \lim_{m\to\infty} \xi_m = y$, $r(m) > 0$ such that

$$\lim_{m\to\infty} \phi^{r(m)}(\xi_m) = y.$$

Let

$$Y_m = \{\phi(\xi_m), \ldots, \phi^{r(m)}(\xi_m)\}.$$

Find $\eta_m \in Y_m$ such that

$$d(\xi_m, \eta_m) = \min_{x\in Y_m} d(x, \xi_m),$$

let

$$\eta_m = \phi^{\rho(m)}(\xi_m).$$

Using techniques of Lemma 0.3.2 find systems ϕ_m such that $\phi_m(\eta_m) = \xi_m$, and

$$\phi_m(x) = \phi(x) \text{ for } x = \xi_m, \ldots, \phi^{\rho(m)-1}(\xi_m),$$

hence $\xi_m \in Per(\phi_m)$. As $\lim_{m\to\infty} d(\xi_m, \eta_m) = 0$ we can find systems ϕ_m so that $\lim_{m\to\infty} \phi_m = \phi$, hence $y \in WPer(\phi)$. It follows from (1.33) - (1.36) that for $\phi \in Z^*$

$$\Omega(\phi) = WPer(\phi) = W\Omega(\phi) = CR(\phi).$$

J.Palis, C.Pugh, M.Shub and D.Sullivan published in [Pa2] a theorem stating that a generic dynamical system $\phi \in Z(M)$ has some very important properties: for example, ϕ has infinitely many periodic points of some fixed period. The proof in [Pa2] is based on the idea of a *permanent periodic orbit* .

Let us say that a periodic orbit γ of ϕ is permanent if any system $\tilde{\phi}$ near ϕ has a periodic orbit $\tilde{\gamma}$ near γ. Denote by perm(ϕ) the union of permanent periodic orbits of $\phi \in Z(M)$. The proof of the main result in [Pa2] is based on the following statement: the map

$$\overline{\mathrm{perm}} : Z(M) \longrightarrow M^*$$

is lower semi-continuous (and hence the set of continuity points of $\overline{\mathrm{perm}}$ is residual in $Z(M)$).

Let us show that there exist dynamical systems ϕ such that the map $\overline{\mathrm{perm}}$ is not lower semi-continuous at ϕ, and systems ϕ such that the map $\overline{\mathrm{perm}}$ is not upper semi-continuous at ϕ. So the proofs in [Pa2] are not complete.

Take $M = S^1$ (with coordinate $\alpha \in [0,1)$). Consider a dynamical system $\phi \in Z(S^1)$ having two fixed points $p = 0, q = \frac{1}{2}$, and such that $\phi(\alpha) > \alpha$ for $\alpha \in (0, \frac{1}{2})$; $\phi(\alpha) < \alpha$ for $\alpha \in (\frac{1}{2}, 1)$. So p is asymptotically stable with respect to ϕ, and q is asymptotically stable with respect to ϕ^{-1}. Obviously, $\overline{\mathrm{perm}}(\phi) = \{0, \frac{1}{2}\}$. Consider a sequence of systems $\phi_m, m = 3, 4, \ldots$, having the following properties:

$$\phi_m(\alpha) = \alpha \text{ for } \alpha \in \{0\} \cup [\frac{1}{2} - \frac{1}{m}, \frac{1}{2} + \frac{1}{m}];$$

$$\phi_m(\alpha) > \alpha \text{ for } \alpha \in (0, \frac{1}{2} - \frac{1}{m}); \phi_m(\alpha) < \alpha \text{ for } \alpha \in (\frac{1}{2} + \frac{1}{m}, 1).$$

We can find systems ϕ_m so that $\lim_{m\to\infty} \rho_0(\phi_m, \phi) = 0$. Fix $m \in \{3, 4, \ldots\}$. It is easy to see that for any $\delta > 0$ there exist systems ψ_1, ψ_2 such that $\rho_0(\phi_m, \psi_i) < \delta, i = 1, 2$, and

$$\mathrm{Per}(\psi_1) = \mathrm{Fix}(\psi_1) = \{0, \frac{1}{2} - \frac{1}{m}\},$$

$$\mathrm{Per}(\psi_2) = \mathrm{Fix}(\psi_2) = \{0, \frac{1}{2} + \frac{1}{m}\}.$$

Hence, for any $r \in [\frac{1}{2} - \frac{1}{m}, \frac{1}{2} + \frac{1}{m}]$ and for any $\delta > 0$ we can find a system ψ such that

$$\rho_0(\phi_m, \psi) < \delta, \text{ and } d(r, \mathrm{Per}(\psi)) \geq \frac{1}{m},$$

so $\overline{\mathrm{perm}}(\phi_m) = \{0\}$. Consequently, the map $\overline{\mathrm{perm}} : Z(S^1) \longrightarrow (S^1)^*$ is not lower semi-continuous at ϕ.

Now take $\phi \in Z(S^1)$ such that $\phi(\alpha) = \alpha$ for $\alpha \in [0,1)$. Evidently, $\overline{\mathrm{perm}}(\phi) = \emptyset$. It is easy to see that we can construct a sequence of systems ϕ_m with $\lim_{m\to\infty} \rho_0(\phi_m, \phi) = 0$ such that for any m the point $p = 0$ is an asymptotically stable fixed point of ϕ_m, hence $\overline{\mathrm{perm}}(\phi_m) \ni \{0\}$. So, $\overline{\mathrm{perm}}$ is not upper semi-continuous at ϕ.

E.Coven, J.Madden and Z.Nitecki corrected in [Cov] the proof of [Pa2] to show that C^0-generically $\Omega(\phi) = \overline{\mathrm{Per}(\phi)}$. Following [Cov] we prove

Theorem 1.4.2. *For a generic dynamical system $\phi \in CLD(M)$*

$$\Omega(\phi) = \overline{Per(\phi)}. \tag{1.37}$$

To prove this theorem we need the following

Lemma 1.4.1. *There exists a residual subset θ of CLD(M) such that for $\phi \in \theta$*

$$Per(\phi) = perm(\phi). \tag{1.38}$$

Proof. We use the notion of fixed point index. Fix $\phi \in Z(M)$. Let U be a an open ball in M whose boundary is an embedded sphere. Suppose that

(1) ϕ has no fixed points on ∂U;

(2) either

(2.a) $\overline{U} \cap \overline{\phi(U)} = \emptyset$, or

(2.b) $\overline{U} \cup \overline{\phi(U)}$ is contained in a single coordinate patch of M.

Assuming that (1) and (2) hold, the fixed point index $\tau(U, \phi)$ is defined in the usual way : if (2.a) holds then $\tau(U, \phi) = 0$; if (2.b) holds then $\tau(U, \phi) = \deg(\gamma)$,where

$$\gamma : \partial U \simeq S^{n-1} \longrightarrow S^{n-1}$$

is defined from local coordinates by

$$\gamma(x) = \frac{\phi(x) - x}{\mid \phi(x) - x \mid},$$

and deg( ) denotes degree in the sense of Hopf [Hi1]. It is easy to understand that this definition is consistent (if (2.a),(2.b) hold) and is independent of the choice of coordinates. It satisfies (see [Hi1])

(3) if $\tau(U, \phi)$ is defined then there exists a neighborhood W of ϕ in $Z(M)$ such that

$$\tau(U, \phi) = \tau(U, \psi)$$

for every $\psi \in W$;

(4) if $\tau(U, \phi) \neq 0$, then ϕ has a fixed point in U ;

(5) if ϕ is a diffeomorphism having exactly one fixed point in U, and this fixed point is hyperbolic, then $\tau(U, \phi) \neq 0$.

Now consider a countable base $\{U_i\}$ for the topology of M, consisting of open balls whose boundaries are embedded spheres. Define for every i two subsets of CLD(M) as follows : $\phi \in A_i$ if

$$\phi(x) \neq x \text{ for } x \in \overline{U_i};$$

$\phi \in B_i$ if there exists U_j such that

$$\text{diam}U_j < \text{diam}U_i, \phi(x) \neq x \text{ for } x \in \partial U_j,$$

$$\tau(U_j, \phi) \neq 0.$$

Evidently, A_i, B_i are open in CLD(M). Hyperbolic fixed points of diffeomorphisms are isolated, hence it follows from the definition of A_i, B_i that if every fixed point of a diffeomorphism ϕ is hyperbolic then $\phi \in A_i \cup B_i$ for any i.

Let KS be the set of Kupka-Smale diffeomorphisms of M. We remind (see [Pi8]) that a diffeomorphism $\phi \in$ KS if :

(a) every periodic point of ϕ is hyperbolic;
(b) stable and unstable manifolds of periodic points of ϕ are transversal.
Now define for $m = 1, 2, \ldots$

$$\theta_{m,i} = \{\phi \in \mathrm{CLD}(M) : \phi^m \in A_i \cup B_i\}.$$

The set KS is a residual subset of $\mathrm{Diff}^1(M)$ (see [Pi8]), hence KS is dense in $\mathrm{CLD}(M)$. It follows from our considerations that

$$\mathrm{KS} \subset \theta_{m,i}$$

for any m, i. Since A_i, B_i are open, and the mapping $\phi \mapsto \phi^m$ is continuous, every $\theta_{m,i}$ is open. So,

$$\theta = \bigcap_{m,i} \theta_{m,i}$$

is a residual subset of $\mathrm{CLD}(M)$. We claim that (1.38) holds for all $\phi \in \theta$. Suppose $\phi \in \theta$ and $\phi^m(x) = x$. Then for every U_i containing x, there exists U_j such that

$$\overline{U_j} \cap U_i \neq \emptyset, \mathrm{diam} U_j < \mathrm{diam} U_i,$$

and $\tau(U_j, \phi^m) \neq 0$. It follows from (3) and (4) that $x \in \mathrm{perm}(\phi)$.

To prove Theorem 1.4.2 note that it follows from Lemma 1.4.1 that the mapping $\phi \mapsto \overline{\mathrm{Per}(\phi)}$ is lower semi-continuous on θ. Apply Lemma 0.2.1 to find a residual subset θ_1 of θ such that any $\phi \in \theta_1$ is a continuity point of the restriction $\overline{\mathrm{Per}}\,|_\theta$. Since $\mathrm{CLD}(M)$ is a Baire space, and θ_1 is a residual subset of θ (and hence of $\mathrm{CLD}(M)$), θ_1 is residual in $\mathrm{CLD}(M)$.

We claim that (1.37) holds for $\phi \in \theta_1$. To get a contradiction suppose that there exists $\phi \in \theta_1$ such that $\Omega(\phi) \neq \overline{\mathrm{Per}(\phi)}$. As for any ϕ

$$\overline{\mathrm{Per}(\phi)} \subset \Omega(\phi),$$

there exists

$$p \in \Omega(\phi) \setminus \overline{\mathrm{Per}(\phi)}.$$

It follows from (1.36) (note that Lemma 0.3.2 holds not only for $Z(M)$ but also for $\mathrm{CLD}(M)$) that there are sequences of systems ϕ_m, and points $p_m \in \mathrm{Per}(\phi_m)$ such that

$$\lim_{m\to\infty} \phi_m = \phi, \lim_{m\to\infty} p_m = p.$$

It is easy to understand that for any m we can find a system ψ_m such that

$$p_m \in \mathrm{perm}(\psi_m), \rho_0(\phi_m, \psi_m) < \frac{1}{m}$$

(for example, this can be done by transforming p_m into a periodic sink for ψ_m).

As θ is dense in $\mathrm{CLD}(M)$ we can find systems $\chi_m \in \theta$ and points $q_m \in \mathrm{Per}(\chi_m)$ such that

$$d(p_m, q_m) < \frac{1}{m}, \rho_0(\chi_m, \psi_m) < \frac{1}{m}.$$

Then

$$\lim_{m\to\infty}\chi_m = \phi, \lim_{m\to\infty} q_m = p,$$

thus we obtain a contradiction with the continuity of $\overline{\text{Per}}\,|_\theta$ at ϕ. This proves Theorem 1.4.2.

As CLD$(M) = Z(M)$ for dim$M \le 3$ the following statement is a corollary of Theorem 1.4.2.

Theorem 1.4.3. *If dimM ≤ 3 then for a generic $\phi \in Z(M)$ (1.37) holds.*

Remark. In the original paper [Cov] the analogue of Theorem 1.4.3 was published without the restriction dim$M \le 3$. But the proof in [Cov] (mostly repeated here) used density of KS in $Z(M)$, and till now this fact was established only in the case dim$M \le 3$ (see Theorem 0.1.1).

Let us now define filtrations for a dynamical system $\phi \in Z(M)$.

Definition 1.14 *A "filtration" for ϕ is a finite collection $\{M_\alpha\}, \alpha \in A = \{0, \dots, m\}$, of compact subsets M_α of M such that:*

(a) $M_0 = \emptyset, M_m = M$;

(b) $M_{\alpha-1} \subset \text{Int}M_\alpha, \alpha = 1, \dots, m$;

(c) $\phi(M_\alpha) \subset \text{Int}M_\alpha, \alpha \in A$.

It follows from (c) that if $\{M_\alpha\}$ is a filtration for $\phi \in Z(M)$ then $\{M_\alpha\}$ is also a filtration for any system ψ in a neighborhood of ϕ in $Z(M)$.

Denote

$$\Lambda_\alpha(\phi) = \bigcap_{k\in\mathbf{Z}} \phi^k(M_\alpha \setminus M_{\alpha-1}), \alpha = 1, \dots, m;$$

$$\Omega_\alpha(\phi) = \Omega(\phi) \cap (M_\alpha \setminus M_{\alpha-1}), \alpha = 1, \dots, m.$$

Lemma 1.4.2. *For any $\alpha \in \{1, \dots, m\}$*

(a) $\Lambda_\alpha(\phi)$ is invariant and compact ;

(b) $\Lambda_\alpha(\phi) \supset \Omega_\alpha(\phi)$.

Proof. Fix α, and let for $n = 1, 2, \dots$

$$B_n = \bigcap_{m=-n}^{n} \phi^m(M_\alpha \setminus M_{\alpha-1}).$$

Evidently $B_{n+1} \subset B_n$, and $\Lambda_\alpha(\phi) = \bigcap_{n\ge1} B_n$. Let us show that for any n

$$\overline{B_{n+1}} \subset B_n. \tag{1.39}$$

Take a sequence $x_k \in B_{n+1}$, and let x be a limit point of this sequence. Any $\phi^m(M_\alpha)$ is compact; $x_k \in \phi^m(M_\alpha)$ for $-n \le m \le n$, hence $x \in \phi^m(M_\alpha), -n \le m \le n$. Suppose that $x \in \phi^m(M_{\alpha-1})$ for some $m \in \{-n, \dots, n\}$. It follows from the property (c) that in this case

$$x \in \phi^{m-1}(\phi(M_{\alpha-1})) \subset \phi^{m-1}(\mathrm{Int} M_{\alpha-1}).$$

Then for large k we have

$$x_k \in \phi^{m-1}(\mathrm{Int} M_{\alpha-1})$$

for some $m-1 \in \{-n-1, \ldots, n\}$, so $x_k \notin B_{n+1}$. The contradiction we obtained proves (1.39). It follows from (1.39) that we can write

$$\Lambda_\alpha(\phi) = \bigcap_{n \geq 1} \overline{B_n},$$

hence $\Lambda_\alpha(\phi)$ is compact. It is evident that $\Lambda_\alpha(\phi)$ is invariant.

To prove (b) take

$$x \in (M_\alpha \setminus M_{\alpha-1}) \setminus \Lambda_\alpha(\phi).$$

We claim that $x \notin \Omega(\phi)$. Indeed, find $m \in \mathbf{Z}$ such that

$$y = \phi^m(x) \notin M_\alpha \setminus M_{\alpha-1}.$$

If $y \in M_{\alpha-1}$ find a neighborhood $U(x)$ of x such that $U(x) \cap \mathrm{Int} M_{\alpha-1} = \emptyset$. For $k > 0$ we have $\phi^k(y) \in \mathrm{Int} M_{\alpha-1}$, hence $x \notin \Omega(\phi)$.

If $y \in M \setminus M_\alpha$ find a neighborhood $U(y)$ of y such that $U(y) \cap \mathrm{Int} M_\alpha = \emptyset$. For $k \geq 1$ we have $\phi^k(x) = \phi^{k-m}(y) \in \mathrm{Int} M_\alpha$, hence $y \notin \Omega(\phi)$ and $x \notin \Omega(\phi)$.

It follows that $\Omega_\alpha(\phi)$ is a subset of $\Lambda_\alpha(\phi)$.

Definition 1.15 *We say that $\{M_\alpha\}$ is a "fine filtration" for ϕ if $\Lambda_\alpha(\phi) = \Omega_\alpha(\phi)$, $\alpha = 1, \ldots, m$.*

S.Smale showed in [Sm3] that if ϕ is a diffeomorphism satisfying Axiom A and the no-cycle condition then ϕ has a fine filtration. We give a proof of this statement in Remark 2 after the proof of Lemma 2.2.7.

For generic systems in $Z(M)$ we can establish the existence of filtrations giving approximations to fine filtrations.

Definition 1.16 *We say that a sequence of filtrations $\{M^k_\alpha\}, k = 1, 2, \ldots; \alpha = 1, \ldots, n_k$, is a "fine sequence" if*

(a) for each $k > 1$ and for each α there is β such that

$$M^k_\alpha \setminus M^k_{\alpha-1} \subset M^{k-1}_\beta \setminus M^{k-1}_{\beta-1};$$

(b) if

$$\Lambda^k_\alpha(\phi) = \bigcap_{m \in \mathbf{Z}} \phi^m(M^k_\alpha \setminus M^k_{\alpha-1}),$$

and

$$\Lambda^k(\phi) = \bigcup_\alpha \Lambda^k_\alpha(\phi)$$

then

$$\bigcap_{k>0} \Lambda^k(\phi) = \Omega(\phi). \tag{1.40}$$

M.Shub and S.Smale proved in [Shu1] two following results.

Theorem 1.4.4. *If a system $\phi \in Z(M)$ has a fine sequence of filtrations then it has no C^0 Ω- explosions.*

Theorem 1.4.5. *If dim$M \geq 2$ and if a system $\phi \in Z(M)$ has no C^0 Ω-explosions then ϕ has a fine sequence of filtrations.*

Remark. [1] It follows from Theorems 1.4.1, 1.4.5 that if dim$M \geq 2$ then a generic system in $Z(M)$ has a fine sequence of filtrations.

Remark. [2] M.Shub and S.Smale considered in [Shu1] dynamical systems generated by diffeomorphisms on a manifold M with dim$M \geq 3$, but their proof is valid also for systems $\phi \in Z(M)$ in the case dim$M \geq 2$. The only place in [Shu1] where the authors use the condition dim$M \geq 3$ is their Lemma 4 which coincides with Lemma 0.3.3 in this book being true for dim$M \geq 2$.

We prove Theorem 1.4.4 below. The reader is referred to the original paper [Shu1] for the proof of Theorem 1.4.5 (it is very useful to take into account the paper [Shu3] where some ideas of [Shu1] are explained in detail).

Proof. (of Theorem 1.4.4) Take arbitrary $\epsilon > 0$. It follows from the property (a) of fine sequences that for each $k > 1, \alpha \in \{1, \dots, n_k\}$ there is β such that

$$\Lambda^k_\alpha(\phi) \subset \Lambda^{k-1}_\beta(\phi),$$

hence $\Lambda^k(\phi) \subset \Lambda^{k-1}(\phi)$. Using (1.38) we can find k such that

$$\Omega(\phi) \subset \Lambda^k(\phi) \subset N_\epsilon(\Omega(\phi)). \tag{1.41}$$

Fix this k. For any $\alpha \in \{1, \dots, n_k\}$ we can write

$$\Lambda^k_\alpha(\phi) = \bigcap_{p\geq 1} \overline{\bigcap_{m=-p}^{p} \phi^m(M^k_\alpha \setminus M^k_{\alpha-1})}$$

(see the proof of Lemma 1.4.2), hence there exists p such that

$$\Lambda^k_\alpha(\phi) \subset \overline{\bigcap_{m=-p}^{p} \phi^m(M^k_\alpha \setminus M^k_{\alpha-1})} \subset N_\epsilon(\Omega(\phi)). \tag{1.42}$$

Evidently we can find p being the same for all α such that (1.42) holds. Then

$$\Lambda^k_\alpha(\phi) \subset \overline{\phi^p(M^k_\alpha) \setminus \phi^{-p}(M^k_{\alpha-1})} \subset N_\epsilon(\Omega(\phi)).$$

Find $\delta > 0$ such that for any system ψ with $\rho_0(\phi, \psi) < \delta$ $\{M_\alpha^k\}$ is a filtration for ψ, and

$$\overline{\psi^p(M_\alpha^k) \setminus \psi^{-p}(M_{\alpha-1}^k)} \subset N_\epsilon(\Omega(\phi))$$

for all α (we remind that we fixed k such that (1.41) is true). For any α

$$\Omega_\alpha(\psi) \subset \Lambda_\alpha^k(\psi) \subset \psi^p(M_\alpha^k) \setminus \psi^{-p}(M_{\alpha-1}^k),$$

so, taking into account that

$$\Omega(\psi) = \bigcup_\alpha \Omega_\alpha(\psi)$$

we obtain finally that

$$\Omega(\psi) \subset N_\epsilon(\Omega(\phi)).$$

This completes the proof.

2. Topological Stability

2.1 General Properties of Topologically Stable Systems

The notion of structural stability is the basic concept in the investigation of the structure of the space $\mathrm{Diff}^1(M)$ of diffeomorphisms of a smooth closed manifold M [Sm2]. The notion of topological stability for homeomorphisms is an analogue of the notion of structural stability. It will be shown later (Theorem 2.2.3) that if a diffeomorphism ϕ is structurally stable then ϕ is also topologically stable.

Various authors used different definitions of topological stability. We use in this book the following type of properties connected with the property of topological stability.

Consider the diagram

$$\begin{array}{ccc} N_1 & \xrightarrow{\psi} & N_2 \\ h\downarrow & & \downarrow h \\ N_3 & \xrightarrow{\phi} & N_4 \end{array} \tag{2.1}$$

where $N_1, \ldots, N_4$ are subsets of M, $\phi, \psi \in Z(M)$, and h is a continuous map.

Definition 2.1 *We say that a dynamical system $\phi \in Z(M)$ is "topologically stable" if given $\epsilon > 0$ there exists a neighborhood W of ϕ in $Z(M)$ such that for any system $\psi \in W$ there is a continuous map h of M onto M having the following properties:*

(a) $d(x, h(x)) < \epsilon$ for $x \in M$;

(b) the diagram (2.1) where $N_i = M$, $i = 1, 2, 3, 4$, is commutative.

This definition of topological stability is close to the corresponding definitions in [Wal, Y2]. In the paper [Ni2] the defined property is called the C^0 lower semi-stability in the strong sense.

Let us describe some connections of topological stability with properties described in Chapter 1.

Theorem 2.1.1 [Moril]. *If ϕ is topologically stable then ϕ is tolerance-$Z(M)$-stable.*

Proof. Fix arbitrary $\epsilon > 0$. Find for ϕ the corresponding neighborhood W in $Z(M)$, and take $\psi \in W$. Write

$$\bar{R}(\Theta(\phi), \Theta(\psi))) = \max(\bar{r}(\Theta(\phi), \Theta(\psi)), \bar{r}(\Theta(\psi), \Theta(\phi))).$$

Consider

$$\bar{r}(\Theta(\phi), \Theta(\psi))) = \max_{A \in \Theta(\phi)} \min_{B \in \Theta(\psi)} R(A, B).$$

Take $A \in \Theta(\phi)$, there exists a sequence of points $x_m \in M$ such that for $A_m = \overline{O(x_m, \phi)}$ we have $\lim_{m\to\infty} R(A_m, A) = 0$.

Let h be a continuous map given by the definition of topological stability. There is a sequence of points $y_m \in M$ such that $h(O(y_m, \psi)) = O(x_m, \phi)$, hence $h(B_m) = A_m$, where $B_m = \overline{O(y_m, \psi)}$. As M^* is compact we can find a convergent subsequence of B_m (let it will be again B_m), and let B be the limit of B_m, so $\lim_{m\to\infty} R(B_m, B) = 0$. It is evident that

$$h(A) = B,$$

it follows from the property (a) of h that

$$R(A, B) \leq \epsilon.$$

As $B \in \Theta(\psi)$ we see that

$$\bar{r}(\Theta(\phi), \Theta(\psi)) \leq \epsilon.$$

Similarly one shows that

$$\bar{r}(\Theta(\psi), \Theta(\phi)) \leq \epsilon,$$

hence

$$\bar{R}(\Theta(\phi), \Theta(\psi)) \leq \epsilon$$

for $\psi \in W$. This completes the proof.

P.Walters [Wa2] and A.Morimoto [Mori1] gave independent proofs of two following results (Theorems 2.1.2, 2.1.3) which describe relations between the topological stability and the POTP (the pseudoorbit tracing property).

Theorem 2.1.2. *If $\phi \in Z(M)$ is topologically stable then ϕ has the POTP.*

Remark. P.Walters and A.Morimoto proved Theorem 2.1.2 in the case $dim M \geq 2$. We are following [Wa2] giving below the proof in this case. The case $M = S^1$ is considered separately.

Let us establish some preliminary results.

Lemma 2.1.1. *Assume that ϕ has the following tracing property for finite pseudoorbits: given $\epsilon > 0$ there exists $\delta > 0$ such that if for a set $\{x_0, \dots, x_m\}$ (1.3) holds for $0 \leq k \leq m-1$ then there exists $x \in M$ with $d(\phi^k(x), x_k) < \epsilon$ for $0 \leq k \leq m-1$. Then ϕ has the POTP.*

Proof. Let $\epsilon > 0$ be given. Choose δ as in the statement of the lemma. Let $\{x_k : k \in \mathbf{Z}\}$ be a δ-trajectory for ϕ. For each $m > 0$ there is $\xi_m \in M$ with

$$d(\phi^k(\xi_m), x_{k-m}) < \epsilon,\ 0 \le k \le 2m.$$

Let $w_m = \phi^m(\xi_m)$. Then

$$d(\phi^k(w_m), x_k) < \epsilon,\ |k| \le m.$$

Choose a convergent subsequence w_m with $\lim_{m\to\infty} w_m = w$. Then

$$d(\phi^k(w), x_k) \le \epsilon,\ k \in \mathbf{Z},$$

and therefore the δ-trajectory $\{x_k\}$ is 2ϵ-traced by $O(w, \phi)$.

Definition 2.2 *We say that two dynamical systems $\phi, \psi \in Z(M)$ are "topologically conjugate" if there exists a homeomorphism $H : M \to M$ such that $\phi \circ H = H \circ \phi$.*

The following statement is "folklore".

Lemma 2.1.2. *Assume that systems $\phi, \psi \in Z(M)$ are topologically conjugate. Then*

(a) ϕ has the POTP if and only if ψ has;

(b) ϕ is topologically stable if and only if ψ is.

Proof. Let us begin by proving the first statement. Let H be a topological conjugacy between ϕ, ψ, and assume that ϕ has the POTP.

Fix arbitrary $\epsilon > 0$. Find $\epsilon_1 > 0$ such that $d(x, y) < \epsilon_1$, $x, y \in M$, implies $d(H^{-1}(x), H^{-1}(y)) < \epsilon$. Now find $\Delta > 0$ such that any Δ-trajectory $\{\xi_k\}$ for ϕ is ϵ_1-traced by a trajectory of ϕ. Finally find $\delta > 0$ such that $d(x, y) < \delta$ implies $d(H(x), H(y)) < \Delta$.

Consider a δ-trajectory $\{x_k\}$ for ψ. Let for $k \in \mathbf{Z}$ $\xi_k = H(x_k)$. As for $k \in \mathbf{Z}$ $d(x_{k+1}, \psi(x_k)) < \delta$ and $\phi \circ H = H \circ \psi$ we have

$$d(\xi_{k+1}, \phi(\xi_k)) = d(H(x_{k+1}), \phi(H(x_k))) =$$
$$= d(H(x_{k+1}), H(\psi(x_k))) < \Delta,$$

so that $\{\xi_k\}$ is a Δ-trajectory for ϕ. Hence, there exists ξ such that

$$d(\xi_k, \phi^k(\xi)) < \epsilon_1, k \in \mathbf{Z}.$$

Let $x = H^{-1}(\xi)$, then for any k

$$d(x_k, \psi^k(x)) = d(H^{-1}(\xi_k), \psi^k(H^{-1}(\xi))) =$$
$$= d(H^{-1}(\xi_k), H^{-1}(\phi^k(\xi))) < \epsilon$$

(we take into account here that $\phi \circ H = H \circ \psi$ implies $H^{-1} \circ \phi = \psi \circ H^{-1}$ and $H^{-1} \circ \phi^k = \psi^k \circ H^{-1}$ for any k). So the first statement is proved.

Now assume that ϕ is topologically stable, that is given $\epsilon > 0$ there exists $\delta > 0$ such that for any $\tilde{\phi} \in Z(M)$ with $\rho_0(\phi, \tilde{\phi})$ there is a continuous map h

mapping M onto M and such that $h \circ \tilde{\phi} = \phi \circ h$. Consider a system ψ and a homeomorphism $H : M \to M$ such that $\phi \circ H = H \circ \psi$.

Let $\tilde{\psi} \in Z(M)$ be a perturbation of ψ. Take $\tilde{\phi} = H \circ \tilde{\psi} \circ H^{-1}$ as a perturbation of ϕ and find the corresponding map h. Define $h_1 = H^{-1} \circ h \circ H$, evidently h_1 is continuous and maps M onto M. Consider

$$h_1 \circ \tilde{\psi} = H^{-1} \circ h \circ H \circ H^{-1} \circ \tilde{\phi} \circ H = H^{-1} \circ h \circ \tilde{\phi} \circ H =$$

$$= H^{-1} \circ \phi \circ h \circ H = H^{-1} \circ H \circ \psi \circ H^{-1} \circ h \circ H = \psi \circ h_1.$$

The rest of the proof is similar to the proof of the first statement.

Let us now prove Theorem 2.1.2. We begin considering the case $\dim M \geq 2$.

Fix arbitrary $\epsilon > 0$. Find $\Delta \in (0, \epsilon)$ such that for any system $\psi \in Z(M)$ with $\rho_0(\phi, \psi) < \Delta$ there exists a map $h : M \to M$ having the properties described in the definition of topological stability. Find $\delta > 0$ such that the statement of Lemma 1.2.2 is true.

Consider a finite set of points $\{x_0, \ldots, x_m\}$ such that (1.3) holds for $0 \leq k \leq m-1$. Apply Lemma 1.2.2 to find a dynamical system ψ and a point ξ such that $\rho_0(\phi, \psi) < \Delta$ and

$$d(\psi^k(\xi), x_k) < \Delta, k = 0, \ldots, m.$$

Let $x = h(\xi)$, then $\phi^k(x) = h(\psi^k(\xi)), k \in \mathbf{Z}$. As

$$d(\phi^k(x), \psi^k(\xi)) < \epsilon, k \in \mathbf{Z},$$

we see that

$$d(\phi^k(x), x_k) \leq \epsilon, k = 0, \ldots, m.$$

As $\epsilon > 0$ is arbitrary it follows from Lemma 1.2.2 that ϕ has the POTP.

Now consider the case $M = S^1$. We apply in this case two results which are proved later:

– a dynamical system $\phi \in Z(S^1)$ is topologically stable if and only if ϕ is topologically conjugate to a Morse-Smale diffeomorphism (see Theorem 2.3.1);

– any Morse-Smale diffeomorphism has the POTP (see Theorem A.1).

These two results and Lemma 2.1.2 imply Theorem 2.1.2 in the case $M = S^1$.

Let us now introduce the property of expansiveness for a dynamical system $\phi \in Z(M)$.

Definition 2.3 *We say that ϕ is "expansive" if there exists $\alpha > 0$ such that if $d(\phi^k(x), \phi^k(y)) \leq \alpha$ for $k \in \mathbf{Z}$ then $x = y$. We say in this case that α is an expansive constant for ϕ.*

Following [Wa2] we now show that if an expansive dynamical system ϕ has the POTP then ϕ is topologically stable. To be exact, the following statement is proved.

Theorem 2.1.3. *Let $\phi \in Z(M)$ be expansive with an expansive constant α. If ϕ has the POTP then given $\epsilon > 0$ with $3\epsilon < \alpha$ there exists $\delta > 0$ such that if $\psi \in Z(M)$ and $\rho_0(\phi, \psi) < \delta$ then there is a unique continuous map $h : M \to M$ such that $h \circ \psi = \phi \circ h$ and $d(x, h(x)) < \epsilon$ for all $x \in M$. If ϵ is small enough then h maps M onto M.*

We begin the proof by establishing two following lemmas.

Lemma 2.1.3. *Let $\phi \in Z(M)$ be expansive with an expansive constant α. Assume that ϕ has the POTP. If $\epsilon \in (0, \alpha/2)$ and δ corresponds to ϵ as in the definition of the POTP then any δ-trajectory is ϵ-traced by a unique trajectory of ϕ.*

Proof. If a δ-trajectory $\{x_k : k \in \mathbf{Z}\}$ is ϵ-traced by trajectories $O(x, \phi), O(y, \phi)$ then for any $k \in \mathbf{Z}$

$$d(\phi^k(x), \phi^k(y)) \leq d(\phi^k(x), x_k) + d(x_k, \phi^k(y)) <$$

$$< 2\epsilon < \alpha,$$

hence $x = y$.

Lemma 2.1.4. *Let $\phi \in Z(M)$ be expansive with an expansive constant α. Given $\lambda > 0$ there exists $N \geq 1$ such that $d(\phi^k(x), \phi^k(y)) \leq \alpha$ for all k with $|k| < N$ implies $d(x, y) < \lambda$.*

Proof. Let $\lambda > 0$ be given. If no N can be chosen with the property stated then for each $N \geq 1$ there exist x_k, y_k with $d(\phi^k(x_k), \phi^k(y_k)) \leq \alpha$ for all k with $|k| < N$ and $d(x_k, y_k) \geq \lambda$. Choose subsequences k_i with $x_{k_i} \to x, y_{k_i} \to y$ as $i \to \infty$. Then $d(x, y) \geq \lambda$, and $d(\phi^k(x), \phi^k(y)) \leq \alpha$ for all k. This contradicts the expansive property of ϕ.

To prove Theorem 2.1.3 fix $\epsilon > 0$ such that $3\epsilon < \alpha$ and choose δ to correspond to ϵ as in the definition of the POTP. Take $\psi \in Z(M)$ such that $\rho_0(\phi, \psi) < \delta$. Then any trajectory of ψ is a δ-trajectory of ϕ. It follows from Lemma 2.1.3 that for any $x \in M$ there exists a unique point $h(x)$ such that $O(x, \psi)$ is ϵ-traced by $O(h(x), \phi)$. This defines a map $h : M \to M$ with

$$d(\phi^k(h(x)), \psi^k(x)) < \epsilon$$

for any $x \in M, k \in \mathbf{Z}$. Putting $k = 0$ gives $d(x, h(x)) < \epsilon$. Since

$$d(\phi^k(h(\psi(x))), \psi^{k+1}(x)) < \epsilon, k \in \mathbf{Z},$$

and

$$d(\phi^k(\phi(h(x))), \psi^{k+1}(x)) = d(\phi^{k+1}(h(x)), \psi^{k+1}(x)) < \epsilon,$$

$k \in \mathbf{Z}$, we have that both $O(h(\psi(x)), \phi)$ and $O(\phi(h(x)), \phi)$ ϵ-trace $\{\psi^{k+1}(x) : k \in \mathbf{Z}\}$. By Lemma 2.1.3 we have $h \circ \psi = \phi \circ h$.

We now show that h is continuous. Let $\lambda > 0$ be given. Using Lemma 2.1.4 choose N so that $d(\phi^k(u), \phi^k(v)) < \alpha$ for $|k| \leq N$ implies $d(u, v) < \lambda$. Choose $\eta > 0$ such that $d(x, y) < \eta$ implies $d(\psi^k(x), \psi^k(y)) < \frac{\alpha}{3}$ for $|k| \leq N$. Then if $d(x, y) < \eta$ we have

$$d(\phi^k(h(x)), \phi^k(h(y))) = d(h(\psi^k(x)), h(\psi^k(y))) \leq$$

$$\leq d(h(\psi^k(x)), \psi^k(x)) + d(\psi^k(x), \psi^k(y)) +$$

$$+ d(\psi^k(y), h(\psi^k(y))) < \epsilon + \frac{\alpha}{3} + \epsilon < \alpha$$

for $|k| \leq N$. Therefore $d(x, y) < \eta$ implies $d(h(x), h(y)) < \lambda$, and the continuity of h is proved.

The map h is the only one with $h \circ \psi = \phi \circ h$ and $d(x, h(x)) < \epsilon$ since if l is another one,

$$d(\phi^k(l(x)), \phi^k(h(x))) = d(l(\psi^k(x)), h(\psi^k(x))) \leq$$

$$\leq d(l(\psi^k(x)), \psi^k(x)) + d(\psi^k(x), h(\psi^k(x))) <$$

$$< 2\epsilon < \alpha$$

for any $k \in \mathbf{Z}$, and so $l(x) = h(x)$.

It is well-known from basic courses of topology that if M is a compact manifold and $d(x, h(x)) < \epsilon$ for any $x \in M$ with ϵ small enough then h maps M onto M. This completes the proof.

2.2 Topological Stability of Systems with Hyperbolic Structure

The notion of a hyperbolic set of a diffeomorphism appeared as one of the most important notions in the theory of structural stability. The topological stability of Anosov diffeomorphisms (that is diffeomorphisms for which the manifold M is a hyperbolic set) was established originally by D.Anosov himself - he wrote about this result in a letter to S.Smale [Ni2]. The first published proof of topological stability of Anosov diffeomorphisms belongs to P.Walters [Wal].

We describe below results of Z.Nitecki [Ni2] who obtained analogues of Smale's and Robbin's results (about Ω-stability and structural stability, respectively) for C^0-perturbations.

We begin by establishing some stability results for a hyperbolic set (see Sect. 0.4 for all definitions).

Let I be a compact invariant set for a diffeomorphism $\phi : M \to M$ of class C^1.

Definition 2.4 *We say that the set I is "locally topologically stable" if there is a neighborhood U of I in M having the following property: given $\epsilon > 0$ there*

exists a neighborhood W of ϕ such that for any system $\psi \in W$ there is an open set V and a continuous map $h : V \to U$ with the following properties:
(a) $d(x, h(x)) < \epsilon$ for $x \in V$;
(b) the diagram (2.1) where $N_1 = V \cap \psi^{-1}(V)$, $N_2 = V, N_3 = U \cap \phi^{-1}(U), N_4 = U$, is commutative.

Remark. This definition corresponds to the definition of the C^0 lower semi-stability of the germ of I in [Ni2].

Theorem 2.2.1 [Ni2]. *A hyperbolic set I of a diffeomorphism $\phi : M \to M$ of class C^1 is locally topologically stable.*

To prove this result we gather some techinical facts we need in the proof.

The space $C^0(M)$ of continuous maps $M \to M$, with the topology of uniform convergence, can be thought as a smooth manifold modelled on the separable Banach space $\mathcal{X}^0(M)$ of continuous vector fields in M. Coordinates at the identity map of M, $id_M \in C^0(M)$, can be given by the exponential map

$$Exp : \mathcal{X}^0(M) \to C^0(M)$$

defined, using the Riemannian metric d on M, by

$$Exp(\xi) : x \mapsto exp_x(\xi_x).$$

Coordinates at other points of $C^0(M)$ can be obtained using pullbacks of the tangent bundle. However, as we shall be working only with neighborhoods of id_M, the above coordinates will suffice for "linearizing" our problems.

We shall use id_M to denote the identity map of M, an element of $C^0(M)$ (and the Exp of the zero section, $O \in \mathcal{X}^0(M)$); we shall use 1 to denote the identity operator on $C^0(M)$.

A diffeomorphism $\phi \in \text{Diff}^1(M)$ gives rise to a differentiable map

$$ad(\phi) : C^0(M) \to C^0(M)$$

defined by

$$ad(\phi)(f) = \phi^{-1} \circ f \circ \phi.$$

This has id_M as a fixed point, and hence can be expressed in local coordinates by

$$Ad(\phi) : \mathcal{X}^0_\epsilon \to \mathcal{X}^0_\epsilon,$$

$$Ad(\phi) = Exp^{-1} \circ ad(\phi) \circ Exp,$$

where

$$[Ad(\phi)(\xi)]_x = exp_x^{-1} \circ \phi^{-1} \circ exp_{\phi(x)}(\xi_{\phi(x)}),$$

and $\mathcal{X}^0_\epsilon$ is the ball of radius $\epsilon > 0$ about the origin in $\mathcal{X}^0$. The derivative at id_M of $Ad(\phi)$ is the invertible linear map

$$\phi_q : \mathcal{X}^0(M) \to \mathcal{X}^0(M)$$

defined by

$$[\phi_q(\xi)]_x = D\phi^{-1}(\xi_{\phi(x)}).$$

Thus, by a standard calculation we can write

$$Ad(\phi)(\xi) = \phi_q(\xi) + Q(\xi) \tag{2.2}$$

where $Q\,|_{\mathcal{X}^0_\epsilon}$ has a small Lipschitz constant when ϵ is small.

For a fixed map $h \in C^0(M)$, we define an operator $\omega_h(f) : C^0(M) \to C^0(M)$ by

$$\omega_h(f) = h \circ f.$$

If h is a C^0-perturbation of id_M, then ω_h takes a neighborhood of id_M into another neighborhood of id_M, and thus can be expressed in local coordinates by

$$\tilde{\omega_h} = Exp^{-1} \circ \omega_h \circ Exp : \mathcal{X}^0_\epsilon \to \mathcal{X}^0_\delta,$$

where

$$[\tilde{\omega_h}(\xi)]_x = exp_x^{-1} \circ h \circ exp_x(\xi_x).$$

Proving Theorem 2.2.1 we shall work with vector fields defined only on compact neighborhoods of a hyperbolic set, so we define some more operators. If U is a compact subset of M then the space $\mathcal{X}^0(U)$ of continuous vector fields defined on U can be given the structure of a Banach space using the usual "sup" norm. If $V \subset U$ are two such compact sets, define

$$\rho_V^U : \mathcal{X}^0(U) \to \mathcal{X}^0(V)$$

by

$$\rho_V^U(\xi) = \xi\,|_V\,.$$

It is clear that ρ_V^U is a norm-decreasing linear map between Banach spaces.

Lemma 2.2.1. *Given $V \subset IntU$, with V, U compact subsets of M, there exists a continuous norm-decreasing linear map*

$$\mathcal{E} : \mathcal{X}^0(U) \to \mathcal{X}^0(M)$$

commuting the diagram

$$\begin{array}{ccc} \mathcal{X}^0(U) & \xrightarrow{\mathcal{E}} & \mathcal{X}^0(M) \\ \rho_V^U \searrow & & \swarrow \rho_V^M \\ & \mathcal{X}^0(V) & \end{array}$$

Proof. Fix a "bump function" Θ subordinate to U, that is a continuous function $\Theta : M \to [0,1]$ with

(a) $supp\Theta \subset U$;

(b) $\Theta(x) = 1$ for $x \in V$.
We define $\mathcal{E}$ by

$$[\mathcal{E}(\xi)]_x = \begin{cases} \Theta(x)\xi_x, & \text{if } x \in U, \\ 0, & \text{if } x \notin U. \end{cases}$$

It is clear that $\mathcal{E}(\xi)$ is continuous, and that $\mathcal{E}$ is linear and norm-decreasing. If $x \in V$ then

$$\mathcal{E}(\xi)_x = \Theta(x)\xi_x = \xi_x,$$

this proves the lemma.

When dealing with locally defined vector fields, in general a neighborhood of an invariant set of ϕ is not mapped into itself by either ϕ or ϕ^{-1}, so that ϕ_q does not define a map of $\mathcal{X}^0(U)$ into itself. However, we can use restrictions and semi-extensions to define some related operators. If we fix $\phi \in \mathrm{Diff}^1(M)$ and U, V, W are compact subsets of M satisfying

$$\phi^{-1}(W) \cup W \cup \phi(W) \subset \mathrm{Int}V \subset V \subset \mathrm{Int}U$$

then we can define two operators

$$\phi_*, \phi^* : \mathcal{X}^0(U) \to \mathcal{X}^0(U)$$

by

$$\phi_* = \rho_U^M \circ \phi_q \circ \mathcal{E}, \phi^* = \rho_U^M \circ \phi_q^{-1} \circ \mathcal{E}$$

where $\mathcal{E}$ is as in Lemma 2.2.1 for U, V. In other words, ϕ_*, ϕ^* can be written

$$[\phi_*(\xi)]_x = D\phi^{-1}[\Theta(\phi(x))\xi_{\phi(x)}] \tag{2.3}$$

$$[\phi^*(\xi)]_x = D\phi[\Theta(\phi^{-1}(x))\xi_{\phi^{-1}(x)}] \tag{2.4}$$

It should be pointed out that neither ϕ_* nor ϕ^* are invertible, but we do have the following weak form of invertibility.

Lemma 2.2.2.

$$\rho_W^U \circ \phi^* \circ \phi_* = \rho_W^U = \rho_W^U \circ \phi_* \circ \phi^*.$$

Proof. We shall prove the first equality; the proof of the second is similar. Suppose $X, Y \in \mathcal{X}^0(U)$. Then by (2.4)

$$[\phi^* Y]_y = D\phi[\Theta(\phi^{-1}(y))Y_{\phi^{-1}(y)}],$$

while by (2.3)

$$[\phi^* X]_{\phi^{-1}(y)} = D\phi^{-1}[\Theta(y)X_y].$$

By substituting $Y = \phi_* X$,

$$[\phi^* \circ \phi_*(X)]_x = D\phi[\Theta(\phi^{-1}(x))D\phi^{-1}(\Theta(x)X_x)].$$

If $x \in W$, then $x, \phi^{-1}(x) \in V$, so that both "Θ" terms are 1, and

$$[(\phi^* \circ \phi_*)(X)]_x = D\phi[D\phi^{-1}(X_x)] = X_x.$$

Since this is true for all $x \in W$, the first equality is proved.

Let

$$TM\mid_I = E^s \oplus E^u \tag{2.5}$$

be the hyperbolic structure on I with constant of hyperbolicity λ (see section 0.4), so that

$$\|D\phi\mid_{E^s_x}\| < \lambda, \|D\phi^{-1}\mid_{E^u_x}\| < \lambda$$

for $x \in I$.

We shall use without a proof the following well-known result of M.Hirsch, J.Palis, C.Pugh and M.Shub.

Lemma 2.2.3 [Hi2]. *If I is a hyperbolic set for a diffeomorphism ϕ with the hyperbolic structure (2.5) then there exists a neighborhood W of I and $\tilde{\lambda} \in (\lambda, 1)$ such that $TM\mid_W$ splits into a continuous Whitney sum of subbundles*

$$TM\mid_W = E^1 \oplus E^2$$

with the following properties:

(a) $D\phi[E^i\mid_{W\cap\phi^{-1}(W)}] = E^i\mid_{\phi(W)\cap W},\ i = 1, 2;$

(b) $\|D\phi\mid_{E^1_x}\| < \tilde{\lambda}, \|D\phi^{-1}\mid_{E^2_x}\| < \tilde{\lambda}, x \in W;$

(c) $E^1_x = E^s_x, E^2_x = E^u_x$ *for* $x \in I$.

Lemma 2.2.4. *Suppose we are given $U \subset M$, the compact closure of an open set, a homeomorphism $f : U \to M$ onto its image uniformly near the identity, and a continuous splitting*

$$TM\mid_{U\cup f^{-1}(U)} = E^1 \oplus E^2.$$

Then, letting

$$\Gamma^i = \{\xi \in \mathcal{X}^0(U) : \xi_x \in E^i_x, x \in U\}, i = 1, 2,$$

and denoting the local representative of α_f by

$$\tilde{\alpha}_f = Exp^{-1} \circ \alpha_f \circ Exp : \mathcal{X}^0(U) \to \mathcal{X}^0(f^{-1}(U))$$

we assert that α_f can be written

$$\tilde{\alpha}_f = J_f + t_f,$$

where

(a) *J_f is linear, preserves Γ^i and is invertible;*

(b) $t_f \mid_{\mathcal{X}^0_\epsilon(U)}$ *(for fixed, small $\epsilon > 0$) is Lipschitz;*
(c) $\|J_f\|, \|J_{f^{-1}}\| \to 1$ *as*

$$\max_{x \in U} d(x, f(x)) \to 0, \tag{2.6}$$

and Lipschitz constant of t_f can be done arbitrary small by taking ϵ and $\max_{x\in U} d(x, f(x))$ small enough.

Proof. The expression for $\tilde{\alpha}_f$ is

$$[\tilde{\alpha}_f(X)]_x = exp_x^{-1}(exp_{f(x)} X_{f(x)}). \tag{2.7}$$

If we let

$$G_{(x,y)} : T_y M \to T_x M$$

be the linear isomorphism defined for $(x, y) \in M \times M$ in a neighborhood of the diagonal by

$$G_{(x,y)} = D(exp_x^{-1} \circ exp_y) \mid_0$$

then it follows easily from (2.7) that

$$[(D\tilde{\alpha}_f)_0(X)]_x = G_{(x,f(x))}[X_{f(x)}].$$

Now, with respect to the splittings,

$$T_y M = E_y^1 \oplus E_y^2, T_x M = E_x^1 \oplus E_x^2,$$

$G_{(x,y)}$ can be expressed by the matrix of linear maps

$$G_{(x,y)} = \begin{pmatrix} A & B \\ C & D \end{pmatrix}_{(x,y)}$$

If we define J_f by

$$[J_f(X)]_x = \begin{pmatrix} A & 0 \\ 0 & D \end{pmatrix}_{(x,f(x))} (X_{f(x)})$$

then J_f clearly satisfies (a),(c) as (2.6) holds.

To show that $t_f = \tilde{\alpha}_f - J_f$ satisfies (b),(c) note that $\|B\|, \|C\| \to 0$ as $d(x, y) \to 0$, so that

$$\left\| \begin{pmatrix} A & B \\ C & D \end{pmatrix}_{(x,f(x))} - \begin{pmatrix} A & 0 \\ 0 & D \end{pmatrix}_{(x,f(x))} \right\| \to 0$$

as (2.6) holds. Furthermore, from the definition of derivative, we have that given $\Delta > 0$ we can find $\epsilon > 0$ such that

$$\|\tilde{\alpha}_f(X)_{f(\cdot)} - G_{(\cdot,f(\cdot))}(X)_{f(\cdot)}\| < \Delta \|X\|$$

when $\|X\| < \epsilon$. Hence $t_f \mid_{\mathcal{X}'_\epsilon(\mathcal{U})}$ is Lipschitz and its Lipschitz constant can be done arbitrary small by taking ϵ and

$$\max_{x \in U} d(x, f(x))$$

small enough.

Let us now prove Theorem 2.2.1. Take a system ψ which is C^0-close to ϕ. We want to find a continuous map h such that the diagram (2.1) is commutative, that is $h \circ \psi = \phi \circ h$ or

$$h \circ \psi \circ \phi^{-1} = \phi \circ h \circ \phi^{-1} \tag{2.8}$$

(we describe below where we solve this equation).

Using our notation we can write (2.8) as

$$\alpha_f(h) = ad(\phi^{-1})(h), \tag{2.9}$$

here $f = \psi \circ \phi^{-1}$. In local coordinates we can express (2.9) as the vector field equation

$$\tilde{\alpha}_f(\eta) = Ad(\phi^{-1})(\eta)$$

or, using the invertibility of $Ad(\phi^{-1}) = [Ad(\phi)]^{-1}$,

$$[1 - Ad(\phi) \circ \tilde{\alpha}_f](\eta) = 0. \tag{2.10}$$

Now find a compact neighborhood U of I (that is the closure of a neighborhood) such that on some open set

$$V \supset \phi^{-1}(U) \cup U \cup \phi(U)$$

the diffeomorphism ϕ has a local hyperbolic structure described by Lemma 2.2.3.

There exists a neighborhood N_0 of ϕ in $Z(M)$ such that for $\psi \in N_0$

$$f(U) \cup f^{-1}(U) \cup \psi(U) \cup \psi^{-1}(U) \subset V.$$

Pick a sequence of compact neighborhoods of I satisfying

$$Z \subset \bigcup_{k=-1}^{1} \phi^k(Z) \subset \mathrm{Int} W \subset \bigcup_{k=-1}^{1} \phi^k(W) \subset$$

$$\subset \mathrm{Int} V \subset V \subset \mathrm{Int} U.$$

Using a semi-extension operator $\mathcal{E}$ as in Lemma 2.2.1 we can look for a solution $\eta \in \mathcal{X}^0(U)$ (with small $\epsilon > 0$) of the well-posed problem

$$\rho_Z^U[1 - \{\rho_U^M \circ Ad(\phi) \circ \tilde{\alpha}_f \circ \mathcal{E}\}](\eta) = \imath. \tag{2.11}$$

The splitting given by the local hyperbolic structure on U is not global so that Lemma 2.2.4 does not apply directly to $\tilde{\alpha}_f : \mathcal{X}^0(M) \to \mathcal{X}^0(M)$, but we are saved by the fact that for any $\eta \in \mathcal{X}^0(U)$ we have $supp[\mathcal{E}(\eta)] \subset U$. So the composition $J_f\mathcal{E}(\eta)$ makes sense. Thus we can write

$$\tilde{\alpha}_f \circ \mathcal{E} = J_f \circ \mathcal{E} + t_f \circ \mathcal{E}$$

where now $J_f \circ \mathcal{E}$ preserves the splitting or else is zero where preservation of the splitting does not make sense. Invoking (2.2) we can rewrite (2.11) as

$$\rho_Z^U[1 - \rho_U^M \phi_q J_f \mathcal{E}](\eta) = \rho_Z^U R_f(\eta) \tag{2.12}$$

where

$$R_f = \rho_U^M Q\tilde{\alpha}_f\mathcal{E} + \rho_U^M \phi_q t_f \mathcal{E}, R_f : \mathcal{X}^0(U) \to \mathcal{X}^0(U).$$

Note that for $\epsilon > 0$ small $R_f \mid_{\mathcal{X}_\epsilon^0(U)}$ is Lipschitz, and we can make its Lipschitz constant arbitrary small by decreasing ϵ and $\rho_0(\phi, \psi)$. This follows from the fact that $Lip(Q) \to 0$ as $\epsilon \to 0$, $\rho_U^M, \tilde{\alpha}_f, \mathcal{E}$, and ϕ_q are Lipschitz-bounded, and Lipschitz constant $Lip(t_f) \to 0$ as $\epsilon \to 0, \rho_0(\phi, \psi) \to 0$ (see Lemma 2.2.4).

We now look for a right inverse to the operator on the left of (2.12). Introduce

$$A_* = \rho_U^M \phi_q J_f \mathcal{E} : \mathcal{X}^0(U) \to \mathcal{X}^0(U),$$
$$A^* = \rho_U^M J_f^{-1} \phi_q^{-1} \mathcal{E} : \mathcal{X}^0(U) \to \mathcal{X}^0(U).$$

We remark that A_*, A^* are well-defined by considerations similar to those preceding Lemma 2.2.2.

Take a neighborhood N_1 if ϕ in $Z(M)$ such that $N_1 \subset N_0$, and for $\psi \in N_1$

$$\bigcup_{k=-1}^{1} f^k(W) \cup \bigcup_{k=-1}^{1} \psi^k(W) \subset \mathrm{Int}V.$$

Then we can show by analogy with Lemma 2.2.2 that

$$\rho_W^U A_* A^* = \rho_W^U = \rho_W^U A^* A_*. \tag{2.13}$$

Let us now show that if $\rho_0(\phi, \psi)$ is small enough then $A_* \mid_{\Gamma^2}$, $A^* \mid_{\Gamma^1}$ are contractions. We give a proof only for the first statement leaving the second one to the reader. If $\xi \in \Gamma^2$, then

$$\|A_*(\xi)\| = \sup_{x \in U} \|D\phi^{-1} \circ A \circ \mathcal{E}(\xi) \circ \phi(x)\|,$$

here A is taken from the representation of J_f in the proof of Lemma 2.2.4. Fix $\lambda_1 \in (\tilde{\lambda}, 1)$, it follows from Lemma 2.2.4 that there exists a neighborhood N_2 of ϕ in $Z(M)$ such that $N_2 \subset N_1$ and if $\psi \in N_2$ then

$$\tilde{\lambda}\|A\| < \lambda_1 \text{ for } x \in U.$$

So, we can write for $\psi \in N_2$

$$\|A_*(\xi)\| = \sup_{x \in U \cap \phi^{-1}(U)} |D\phi^{-1}[\Theta(\phi(x))A\xi_{\phi(x)}]| \leq$$

$$\leq \tilde{\lambda} \sup_{x \in U \cap \phi^{-1}(U)} |A\xi_{\phi(x)}| \leq$$

(we take into account that for $\xi \in \Gamma^2$ we have $A\xi \in \Gamma^2$)

$$\leq \tilde{\lambda} \sup_{x \in U} \|A\| \sup_{y \in \phi(U) \cap U} |\xi_y| \leq \lambda_1 \|\xi\|.$$

So, $A_* |_{\Gamma^2}$ is a contraction. We take N_2 such that for $\psi \in N_2$ $A^* |_{\Gamma^1}$ is also a contraction (with the same contraction constant λ_1).

Define the operator

$$K : \mathcal{X}^0(U) \to \mathcal{X}^0(U)$$

by

$$K |_{\Gamma^2} = \sum_{k=0}^{\infty} (A_*)^k |_{\Gamma^2},$$

$$K |_{\Gamma^1} = -\sum_{k=1}^{\infty} (A^*)^k |_{\Gamma^1}.$$

We claim that

$$\rho_W^U [1 - A_*] K = \rho_W^U. \tag{2.14}$$

We prove (2.14) separately on Γ^2 and Γ^1. For $\xi \in \Gamma^2(U)$, clearly

$$[(1 - A_*) \circ K](\xi) = \lim_{m \to \infty} [\xi - (A_*)^m \xi] = \xi,$$

so that application of ρ_W^U to both sides gives (2.14) on Γ^2. Consider

$$\rho_W^U (1 - A_*) \circ K = -\rho_W^U \circ (1 - A_*) \circ A^* \circ \sum_{k=0}^{\infty} (A^*)^k =$$

$$= -\rho_W^U (A^* - A_* A^*) \circ \sum_{k=0}^{\infty} (A^*)^k =$$

$$= -[\rho_W^U A^* - \rho_W^U A_* A^*] \circ \sum_{k=0}^{\infty} (A^*)^k =$$

$$= -[\rho_W^U A^* - \rho_W^U] \circ \sum_{k=0}^{\infty} (A^*)^k =$$

(we take (2.13) into account here)

$$= -\rho_W^U [A^* - 1] \circ \sum_{k=0}^{\infty} (A^*)^k =$$

$$= -\rho_W^U \lim_{m \to \infty} [(A^*)^m - 1] = \rho_W^U$$

establishing (2.14) on Γ^1.

As $A_* |_{\Gamma^2}$, $A^* |_{\Gamma^1}$ are contractions with contraction constant λ_1 we have

$$\|K\| \leq \sum_{k=0}^{\infty}(\lambda_1)^k = \frac{1}{1-\lambda_1}.$$

On the other hand, we can find $\epsilon > 0$ and a neighborhood N_3 of ϕ such that $N_3 \subset N_2$, and for $\psi \in N_3$

$$Lip(R_f) < 1 - \lambda_1.$$

Then, by the Contraction Theorem, $K \circ R_f$ has a fixed point $\eta \in \mathcal{X}^0_\epsilon(U)$:

$$\eta = K \circ R_f(\eta).$$

Composing on the left by $\rho^U_W[1 - A_*]$ gives, in view of (2.14)

$$\rho^U_W[1 - A_*](\eta) = \rho^U_W R_f(\eta).$$

Composing again on the left by ρ^W_Z and noting that $\rho^W_Z \circ \rho^U_W = \rho^U_Z$ we get

$$\rho^U_Z[1 - A_*](\eta) = \rho^U_Z R_f(\eta)$$

which is just (2.12). Thus, the map defined on U by $h = Exp(\eta)$ has the required property on Z. As ϵ is arbitrary small, we can get $\rho_0(h, id)$ arbitrary small, then h will map Z onto a neighborhood of I.

Now we are going to prove one of the basic results obtained by Z.Nitecki in [Ni2] - the theorem giving sufficient conditions for topological Ω-stability and being an analogue of the classical Ω-stability Theorem of S.Smale [Sm3] for C^0-perturbations.

We begin with the following

Definition 2.5 *We say that a system $\phi \in Z(M)$ is "topologically Ω-stable" if given $\epsilon > 0$ there exists a neighborhood W of ϕ in $Z(M)$ such that for any system $\psi \in W$ there is a continuous map h mapping $\Omega(\psi)$ onto $\Omega(\phi)$ and such that*

(a) $d(x, h(x)) < \epsilon$ for $x \in \Omega(\psi)$;

(b) the diagram (2.1) where $N_1 = N_2 = \Omega(\psi)$, $N_3 = N_4 = \Omega(\phi)$ is commutative.

Remark. In [Ni2] the corresponding property of ϕ is called the C^0 lower Ω-semi-stability in the strong sense.

Theorem 2.2.2 [Ni2]. *If ϕ is a diffeomorphism of class C^1 satisfying Axiom A and the no-cycle condition, then ϕ is topologically Ω-stable.*

To prove this result we need Smale's techniques for constructing filtrations [Sm3] (we are following [Ni1] here).

Consider a basic set Ω_l for a diffeomorphism ϕ. For $\epsilon > 0$ we denote by $W^\sigma(x, \epsilon)$ the ϵ-neighborhood of a point $x \in \Omega_l$ in $W^\sigma(x), \sigma = s, u$ (with respect to the inner metrics of $W^u(x), W^s(x)$). Denote

$$W^\sigma(\Omega_l, \epsilon) = \bigcup_{x \in \Omega_l} W^\sigma(x, \epsilon), \sigma = s, u.$$

Let us say that a compact subset $D \subset W^u(\Omega_l, \epsilon)$ is a *fundamental domain* for $W^u(\Omega_l, \epsilon)$, if

$$W^u(\Omega_l, \epsilon) \setminus \Omega_l \subset O^-(D, \phi).$$

We remind that for a set $X \subset M$ and for a system $\phi \in Z(M)$

$$O^-(X, \phi) = \bigcup_{k \le 0} \phi^k(X)$$

and so on.

We say that a fundamental domain D for $W^u(\Omega_l, \epsilon)$ is proper if $D \cap \Omega_l = \emptyset$.

Lemma 2.2.5 [Nil]. *For $\epsilon > 0$ small enough,*

$$D^u = \overline{W^u(\Omega_l, \epsilon) \setminus \phi^{-1}(W^u(\Omega_l, \epsilon))}$$

is a proper fundamental domain for $W^u(\Omega_l, \epsilon)$.

Proof. As ϕ^{-1} contracts $W^u(\Omega_l, \epsilon)$ we have

$$\bigcap_{k \le 0} \phi^k(W^u(\Omega_l, \epsilon)) = \Omega_l.$$

Hence,

$$W^u(\Omega_l, \epsilon) \setminus \Omega_l = W^u(\Omega_l, \epsilon) \setminus \bigcap_{k \le 0} \phi^k(W^u(\Omega_l, \epsilon)) \subset$$
$$\subset O^-(D^u, \phi),$$

so D^u is a fundamental domain. To show that D^u is a proper fundamental domain find $\delta > 0$ such that $0 < \delta < \epsilon$, and

$$W^u(\Omega_l, \delta) < \phi^{-1}(W^u(\Omega_l, \epsilon)).$$

We claim that $W^u(\Omega_l, \delta)$ is a neighborhood of Ω_l in $W^u(\Omega_l, \epsilon)$. To obtain a contradiction suppose that there exist sequences of points $p_m \in \Omega_l$,

$$x_m \in W^u(p_m, \epsilon) \setminus W^u(\Omega_l, \delta)$$

such that $x_m \to p \in \Omega_l$ as $m \to \infty$. It is shown in the proof of Theorem 0.4.2 in [Sm2] that in such a situation if ϵ is small enough then $W^u(p_m, 2\epsilon)$ and $W^s(p, 2\epsilon)$ have a unique point q_m of intersection, and $q_m \in \Omega$ (hence for large m $q_m \in \Omega_l$). As $\lim_{m\to\infty} x_m = p, \lim_{m\to\infty} q_m = p$ we have $\lim_{m\to\infty} d(x_m, q_m) = 0$. Taking into account that $W^u(x_m) = W^u(q_m)$ we see that $x_m \in W^u(q_m, \delta)$ for large m. The contradiction we obtained proves that $\phi^{-1}(W^u(\Omega_l, \epsilon))$ is a neighborhood of Ω_l in $W^u(\Omega_l, \epsilon)$. Therefore, Ω_l cannot intersect the closure of the set

$$W^u(\Omega_l, \epsilon) \setminus \phi^{-1}(W^u(\Omega_l, \epsilon)),$$

so we see that $D^u \cap \Omega_l = \emptyset$.

Lemma 2.2.6 [Nil]. *Let U be a neighborhood of Ω_l. Assume that there exists a compact set $X \subset M$ such that $X \cap \Omega_l = \emptyset$, and*
(a) $\phi(X) \subset \mathrm{Int}X$;
(b) $O^-(X,\phi) \cup \Omega_l \supset W^u(\Omega_l)$.
Then there exists a natural m and a compact set Y, $X \subset Y \subset M$, such that
(a) $\phi(Y) \subset \mathrm{Int}Y$;
(b) $\Omega_l \subset \mathrm{Int}Y \setminus \phi^{-m}(X) \subset U$.

Proof. As basic sets are disjoint compact sets and any trajectory belongs to $W^u(\Omega_j)$ for some j we may assume that U is compact and

$$\bigcap_{k\geq 0} \phi^k(U) \subset W^u(\Omega_l).$$

It follows from Lemma 2.2.5 that there exists a proper fundamental domain D^u for $W^u(\Omega_l,\epsilon)$ for some $\epsilon > 0$ having the following properties:
(1) $O^-(D^u,\phi) \subset \mathrm{Int}U$,
(2) $\mathrm{Int}(\phi^{-m}(X)) \supset D^u$ for some $m > 0$.
Define the following sets

$$P = W^u(\Omega_l) \cup (\bigcap_{k>0} \phi^k(X)),$$

$$Q = U \cup \phi^{-m}(X).$$

We claim that

$$P \subset \bigcap_{k\geq 0} \phi^k(Q). \tag{2.15}$$

As $\phi(X) \subset X$ we have

$$\bigcup_{j\leq m} \phi^{-j}(X) \subset \phi^{-m}(X).$$

Hence $O^+(D^u,\phi) \subset \phi^{-m}(X)$. As $O^-(D^u,\phi) \subset U$ we obtain that

$$O(D^u,\phi) \subset U \cup \phi^{-m}(X) = Q.$$

It follows that for any k $O(D^u,\phi) \subset \phi^k(Q)$, consequently

$$W^u(\Omega_l) \setminus \Omega_l = O(D^u,\phi) \subset \bigcap_{k\geq 0} \phi^k(Q).$$

But $X \subset \phi^{-m}(X) \subset Q$, hence

$$\bigcap_{k\geq 0} \phi^k(X) \subset \bigcap_{k\geq 0} \phi^k(Q),$$

this proves (2.15).

Let us now find a compact set V such that

$$\phi(V) \subset \text{Int}V, P \subset \text{Int}V \subset V \subset Q. \tag{2.16}$$

Define for $r > 0$

$$A_r = \bigcap_{m=0}^{r} \phi^m(Q).$$

Note that $P \subset \text{Int}Q$. Indeed, we can write

$$P = O^+(D^u, \phi) \cup O^-(D^u, \phi) \cup (\bigcap_{k>0} \phi^k(X)),$$

but

$$O^-(D^u, \phi) \subset \text{Int}U \subset \text{Int}Q,$$
$$O^+(D^u, \phi) = D^u \cup O^+(\phi(D^u), \phi) \subset$$
$$\subset \text{Int}\phi^{-m}(X) \cup (\phi^{-m+1}(X)) \subset \text{Int}\phi^{-m}(X),$$

so

$$P \subset \text{Int}Q.$$

As

$$A_1 \supset A_2 \supset \ldots,$$

and

$$\bigcap_{r \geq 0} A_r = P,$$

there exists r such that $\phi(A_r) \subset \text{Int}Q$, hence $\phi(A_r) \subset A_{r+1}$. We can find $m \geq 2$ such that

$$\phi_m(A_r) \subset \text{Int}A_r.$$

Choose now W compact such that

$$A_r \subset \text{Int}W, \phi^m(W) \subset \text{Int}A_r,$$

and let

$$E = A_r \cup (W \cap \phi^{m-1}(W)).$$

We claim that

$$\phi^{m-1}(E) \subset \text{Int}E. \tag{2.17}$$

Indeed,

$$\phi^{m-1}(A_r) \subset \text{Int}\phi^{m-1}(W),$$
$$\phi^{m-1}(A_r) \subset A_{r+1} \subset A_r \subset \text{Int}W,$$

so

$$\phi^{m-1}(A_r) \subset (\text{Int}W) \cap \text{Int}\phi^{m-1}(W) \subset \text{Int}E. \tag{2.18}$$

As $m \geq 2$,

$$\phi^{m-1}(W \cap \phi^{m-1}(W)) \subset \phi^{2m-2}(W) = \phi^{m-2}(\phi^m(W)) \subset$$

$$\subset \phi^{m-2}(\mathrm{Int}A_r) \subset \mathrm{Int}A_r \subset \mathrm{Int}E. \tag{2.19}$$

We see that (2.18),(2.19) imply (2.17).

Repeating this process with E instead of A_r we find E_1 for which an analogue of (2.17) holds with $m-2$ instead of $m-1$ and so on. As a result we obtain a compact neighborhood V of P such that $V \subset Q$, $\phi(V) \subset \mathrm{Int}V$, and

$$\bigcap_{m\geq 0} \phi^m(V) = P.$$

Let now

$$Y = V \cup \phi^{-m}(X).$$

It is evident that $\phi(Y) \subset \mathrm{Int}Y$. As $Y \subset Q$ we have $\mathrm{Int}Y \setminus \phi^{-m}(X) \subset U$. As $\Omega_l \subset P$, we have $\Omega_l \subset \mathrm{Int}V$, hence

$$\Omega_l \subset \mathrm{Int}Y \setminus \phi^{-m}(X).$$

This completes the proof.

It follows from the no-cycle condition that we can choose a simple ordering $<$ on the set of basic sets using indices so that

$$\Omega_1 < \Omega_2 < \ldots < \Omega_p,$$

and if $\Omega_i < \Omega_j$ then it is false that $\Omega_i \to \Omega_j$ (that is if $\Omega_i < \Omega_j$ then

$$W^u(\Omega_i) \cap W^s(\Omega_j) = \emptyset).$$

Apply Theorem 0.4.4 to fix disjoint neighborhoods U_i of the sets $\Omega_j, j = 1,\ldots,p$, such that

$$\bigcap_{m\in \mathbf{Z}} \phi^m(U_j) = \Omega_j,$$

$$\bigcap_{m>0} \phi^m(U_j) = W^u(\Omega_j) \cap U_j,$$

$$\bigcap_{m>0} \phi^{-m}(U_j) = W^s(\Omega_j) \cap U_j.$$

Now we prove the existence of a filtration for ϕ.

Lemma 2.2.7 [Ni1]. *There exist compact subsets M_j of M, $j = 0,\ldots,p$, and natural numbers $m_1,\ldots,m_p$ such that*

$$\emptyset = M_0 \subset M_1 \subset \ldots \subset M_p = M,$$

and for $j = 1,\ldots,p$:
(a) $\phi^{-m_j}(M_{j-1}) \subset M_j$;
(b) $\phi(M_j) \subset \mathrm{Int}M_j$;
(c) $\Omega_j \subset \mathrm{Int}M_j \setminus \phi^{-m_j}(M_{j-1}) \subset U_j$.

Proof. We prove the lemma using induction on j. Consider $j = 1$. Note that

$$W^u(\Omega_1) \cap W^s(\Omega_i) = \emptyset \text{ for } i > 1.$$

Hence

$$M = \bigcup_{i=1}^{p} W^s(\Omega_i) \tag{2.20}$$

(see Theorem 0.4.3) implies

$$W^u(\Omega_1) = \Omega_1.$$

Take $X = \emptyset$ in Lemma 2.2.6, and denote by M_1 the corresponding set Y.

Now suppose that the statement of lemma is true for $j \leq l - 1$. We claim that

$$W^u(\Omega_l) \subset \Omega_l \cup \bigcup_{k>0} \phi^{-k}(M_{l-1}). \tag{2.21}$$

As

$$W^u(\Omega_l) \cap W^s(\Omega_i) = \emptyset \text{ for } i > l$$

it follows from (2.20) that

$$W^u(\Omega_l) = \bigcup_{i=1}^{p} W^u(\Omega_l) \cap W^s(\Omega_i) =$$

$$= \bigcup_{i \leq l} W^u(\Omega_l) \cap W^s(\Omega_i). \tag{2.22}$$

Take $i < l$. As

$$\Omega_i \subset \operatorname{Int} M_i \subset \operatorname{Int} M_{l-1},$$

for some $\delta > 0$

$$W^s(\Omega_i, \delta) \subset M_{l-1}.$$

For any $x \in W^s(\Omega_i)$ there is $k > 0$ such that

$$\phi^k(x) \in W^s(\Omega_i, \delta),$$

hence

$$W^s(\Omega_i) \subset \bigcup_{k>0} \phi^{-k}(M_{l-1}). \tag{2.23}$$

Taking into account that the no-cycle condition implies

$$W^s(\Omega_l) \cap W^u(\Omega_l) = \Omega_l$$

we see that (2.21) is a concequence of (2.22). (2.23).

We can now apply Lemma 2.2.6 taking $X = M_{l-1}$, and denoting $M_l = Y$. Properties (a)–(c) follow from Lemma 2.2.6 and from the description of the induction process.

Remark. [1] Consider neighborhoods W_j of basic sets $\Omega_j, j = 1, \ldots, p$:

$$W_j = U_j \cap \mathrm{Int} M_j.$$

We claim that these neighborhoods are unrevisited, that is if $x \in W_j$, and for some $\kappa > 0$ we have $\phi^\kappa(x) \notin W_j$ then $\phi^k(x) \notin W_j$ for $k \geq \kappa$.

Indeed, it follows from the property (b) of M_j that if $x \in \mathrm{Int} M_j$ then $\phi^k(x) \in \mathrm{Int} M_j$ for all $k \geq 0$. Hence if $y = \phi^\kappa(x) \notin W_j$ then $y \notin U_j$, so that

$$y \in \phi^{-m_j}(M_{j-1})$$

(we refer to the property (c) of M_j). But then for $k \geq 0$ we have

$$\phi^k(y) \subset \phi^{k-m_j}(M_{j-1}) \subset \phi^{-m_j}(M_{j-1}),$$

hence $\phi^{k+\kappa}(x) = \phi^k(y) \notin W_j$.

It is evident that if compact sets $M_1, \dots, M_p$ and a system $\phi \in Z(M)$ have properties (a)-(c) described in Lemma 2.2.7 then there exists a neighborhood N_0 of ϕ in $Z(M)$ such that any system $\psi \in N_0$ has the following property. Letting

$$\Omega_j(\psi) = \Omega(\psi) \cap U_j, j = 1, \dots, p,$$

the properties (a)-(c) hold with ψ substituted for ϕ everywhere. Neighborhoods W_j are unrevisided by trajectories of ψ and

$$\Omega(\psi) = \bigcup_{j=1}^{p} \Omega_j(\psi).$$

Remark. [2] Let us show that the filtration $\{M_j\}$ constructed in Lemma 2.2.7 is a fine filtration for the system ϕ (see Sect. 1.4 for definitions).

It was shown in Lemma 1.4.1 that for $j = 1, \dots, p$ we have

$$\Omega_j(\phi) \subset \Lambda_j(\phi). \tag{2.24}$$

Evidently for any j

$$\Omega_j \subset \Omega_j(\phi). \tag{2.25}$$

We claim that for any j

$$\Lambda_j(\phi) \subset \Omega_j. \tag{2.26}$$

Take $x \in \Lambda_j(\phi) = \bigcap_{m \in \mathbf{Z}} \phi^m(M_j \setminus M_{j-1})$. It was shown in Lemma 1.4.1 that $\Lambda_j(\phi)$ is invariant, so $\phi^k(x) \in \Lambda_j(\phi)$ for any $k \in \mathbf{Z}$. As

$$\phi(M_j) \subset \mathrm{Int} M_j,$$

$$\phi(M_{j-1}) \subset M_{j-1} \subset \phi^{-1}(M_{j-1}) \subset \dots \subset \phi^{-m_j}(M_{j-1}),$$

we see that

$$\Lambda_j(\phi) \subset \phi(M_j) \setminus \phi(M_{j-1}) \subset$$

$$\subset \mathrm{Int} M_j \setminus \phi^{-m_j}(M_{j-1}) \subset U_j,$$

hence $O(x, \phi) \subset U_j$. It follows from the choice if U_j that

$$x \in \Omega_j.$$

So, (2.26) is proved. It follows from (2.24)–(2.26) that for $j = 1, \dots, p$

$$\Omega_j(\phi) = \Lambda_j(\phi),$$

and we see that the filtration $\{M_j\}$ is fine.

Let us now prove Theorem 2.2.2. Using Lemma 2.2.3, we pick disjoint neighborhoods U_j of the basic sets $\Omega_j, j = 1, \dots, p$ of ϕ, on which there is a local hyperbolic structure extending that on Ω_j. Shrinking U_j if necessary we assume by Theorem 2.2.1 that for any system ψ in a neighborhood N_1 if ϕ in $Z(M)$ there exist continuous maps $h_j(\psi), j = 1, \dots, p$ which map U_j onto a neighborhood of $\text{Int} M_j \setminus \phi^{-m_j}(M_{j-1})$, and for $x \in U_j \cap \psi^{-1}(U_j)$ we have

$$h(\psi(x)) = \phi(h(x)).$$

Taking $\psi \in N_0 \cap N_1$ (see Remark 1 after Lemma 2.2.7) so that ψ satisfies analogues of the properties (a)–(c) in this lemma we can define

$$h : \bigcup_{j=1}^{p} U_j \to M$$

by

$$h\,|_{U_j} = h_j(\psi).$$

We claim that

$$h(\Omega(\psi)) = \Omega(\phi). \tag{2.27}$$

Take any point $x \in \Omega_j(\psi), j \in \{1, \dots, p\}$. There exists a sequence of points x_m, and a sequence of numbers k_m such that

$$x_m \to x, k_m \to +\infty, \psi^{k_m}(x_m) \to x \text{ as } m \to \infty.$$

As the neighborhood W_j is unrevisited by trajectories of ψ we see that for large m

$$\psi^k(x_m) \in W_j \text{ for } 0 \le k \le k_m,$$

hence if $y = h(x), y_m = h(x_m)$ then

$$y_m \to y, \phi^{k_m}(y_m) = h(\psi^{k_m}(x_m)) \to y \text{ as } m \to \infty,$$

so that $y \in \Omega(\phi)$. Thus,

$$h(\Omega(\psi)) \subset \Omega(\phi). \tag{2.28}$$

To show that

$$\Omega(\phi) \subset h(\Omega(\psi)) \tag{2.29}$$

it suffices to show that

$$\mathrm{Per}(\phi\,|_{U_j}) \subset h(\Omega_j(\psi)), j = 1, \ldots, p, \tag{2.30}$$

since the periodic points of ϕ are dense in $\Omega(\phi)$, and $\Omega_j(\psi)$ is compact.

Take a periodic point $p \in \mathrm{Per}(\phi\,|_{U_j})$. Its trajectory, $O(p, \phi)$ is a compact ϕ-invariant subset of U_j, in fact of Ω_j. Hence, the inverse image

$$h^{-1}[O(p, \phi)]$$

is a compact ψ- invariant subset of W_j. Since any compact invariant set must contain nonwandering points,

$$h^{-1}[O(p, \phi)] \cap \Omega_j(\psi) \neq \emptyset,$$

$$O(p, \phi) \cap h[\Omega_j(\psi)] \neq \emptyset,$$

is a ϕ- invariant subset of W_j, and hence contains p. Thus, we proved (2.30), and (2.29).

Now (2.28),(2.29) imply (2.27).

Fix arbitrary $\epsilon > 0$. It follows from Theorem 2.2.1 that we can find a neighborhood $N \subset N_0 \cap N_1$ of ϕ such that for any $\phi \in N$ the corresponding map h has the property: $d(x, h(x)) < \epsilon$ for $x \in \Omega(\psi)$.

Now we state the second basic result of the paper [Ni2] – the theorem giving sufficient conditions for the topological stability of a diffeomorphism. This theorem is an analogue of classical structural stability theorems of J.Robbin and C.Robinson [Robb, Robi1] for C^0- perturbations.

Theorem 2.2.3 [Ni2]. *If a diffeomorphism ϕ satisfies the STC then ϕ is topologically stable.*

We refer the reader to the original paper [Ni2] for a proof of this theorem.

2.3 Topologically Stable Dynamical Systems on the Circle

We describe here some results of K.Yano [Y1] who studied topological stability of systems $\phi \in Z(S^1)$.

Consider the circle S^1 with coordinate $\alpha \in [0, 1)$. We identify a system $\phi \in Z(S^1)$ with the corresponding map $\phi : [0, 1) \to [0, 1)$.

In Sect. 0.4 Morse-Smale diffeomorphisms of S^1 were defined. It was mentioned that a diffeomorphism ϕ of S^1 is Morse-Smale if and only if:

(a) $\mathrm{Per}(\phi) \neq \emptyset$;

(b) every trajectory in $\mathrm{Per}(\phi)$ is hyperbolic.

Definition 2.6 *Let ϕ be an orientation preserving homeomorphism of S^1. A fixed point x of ϕ is said to be "topologically hyperbolic" if it is isolated in $\mathrm{Fix}(\phi)$, and $\phi(\alpha) - \alpha$ changes its sign at $\alpha = x$.*

A periodic point x of any homeomorphism g of S^1 is called topologically hyperbolic if x is a topologically hyperbolic fixed point of g^{2n}, where n is a period of x.

Theorem 2.3.1 [Y1]. *A system $\phi \in Z(S^1)$ is topologically stable if and only if ϕ is topologically conjugate to a Morse-Smale diffeomorphism.*

Proof. It follows from Theorem 2.2.3 that any Morse-Smale diffeomorphism is topologically stable. Apply Lemma 2.1.2 to show that if ϕ is topologically conjugate to a Morse-Smale diffeomorphism, then ϕ is topologically stable.

To prove the converse statement note that the following fact is geometrically evident: $\phi \in Z(S^1)$ is topologically conjugate to a Morse-Smale diffeomorphism if and only if ϕ satisfies the following two conditions:

(a′) Per(ϕ) is non-empty and finite;

(b′) every element of Per(ϕ) is topologically hyperbolic.

Hence, to prove the theorem, it suffices to show that every topologically stable system $\phi \in Z(S^1)$ satisfies the above conditions (a'),(b').

First we prove the following

Lemma 2.3.1. *Assume that ϕ, ψ are orientation preserving homeomorphisms of S^1 and that there exists a continuous map h of S^1 onto S^1 such that $h \circ \psi = \phi \circ h$. Then if* Per($\psi$) *is non-empty, so is* Per(ϕ). *There exists a constant C depending only on ϕ such that if $d(\alpha, h(\alpha)) \leq C$ for $\alpha \in S^1$ then*

$$\text{card Per}(\psi) \geq \text{card Per}(\phi).$$

Proof. Since for every $\phi \in Z(S^1)$ there exists an integer m such that Per(ϕ) =Fix(ϕ^m), it is enough to prove the lemma replacing Per by Fix. The proof of the first part is immediate.

To prove the second part, we take a positive C satisfying the following two conditions:

(1) $C \leq \frac{1}{8}$;

(2) if I is a closed interval in S^1 of length not greater than $4C$ then the length of $\phi(I)$ is not greater than $\frac{1}{4}$.

Suppose that for ψ, h we have $h \circ \psi = \phi \circ h$, and $d(\alpha, h(\alpha)) \leq C, \alpha \in S^1$. Take a fixed point x of ϕ. Since $h^{-1}(x)$ is a non-empty closed ψ-invariant subset of S^1 contained in $[x - C, x + C]$, we can define sup $h^{-1}(x)$ and inf $h^{-1}(x)$ without confusion. Put

$$B = [\inf h^{-1}(x), \sup h^{-1}(x)].$$

Then we have

$$(\text{length of B}) \leq 2C,$$

and

$$(\text{length of } \psi(B)) \leq (\text{length of } h(\psi(B)))+$$

$$+2\max_{\alpha\in S^1} d(\alpha,h(\alpha)) \leq (\text{length of } \phi(h(B))) + 2C \leq$$

$$\leq \frac{1}{4}+\frac{1}{4}=\frac{1}{2}.$$

To obtain the last inequality, we used the condition (2), and the estimate

$$(\text{length of } h(B)) \leq (\text{length of } B) + 2\max_{\alpha\in S^1} d(\alpha,h(\alpha)) \leq 4C.$$

Hence

$$(\text{length of } B) + (\text{length of } \psi(B)) \leq 2C + \frac{1}{2} \leq \frac{3}{4}.$$

By the ψ-invariance of $h^{-1}(x)$, both end points of $\psi(B)$ are in B, and those of B are in $\psi(B)$. Because the total length of S^1 is one, the above estimate shows that B is ψ-invariant. Since ψ is orientation preserving, sup $h^{-1}(x)$ is a fixed point of ψ. Hence we have an injection from $\mathrm{Fix}(\phi)$ to $\mathrm{Fix}(\psi)$ which maps x to sup $h^{-1}(x)$. This completes the proof.

Now suppose that $\phi \in Z(S^1)$ is topologically stable. Since every homeomorphism of S^1 can be approximated by a diffeomorphism [Mu1], it follows from a theorem of M.Peixoto (see a proof in [Ni1]) that in any neighborhood of ϕ there exists a Morse-Smale diffeomorphism. So we can find a Morse-Smale diffeomorphism ψ and a continuous map h of S^1 onto S^1 such that $h\circ\psi = \phi\circ h$. Therefore, by Lemma 2.3.1, $\mathrm{Per}(\phi)$ is non-empty and finite (if f is orientation reversing, apply Lemma 2.3.1 to ϕ^2).

Next, take a periodic point x of ϕ. If x is not topologically hyperbolic, we can eliminate this periodic point by arbitrary small perturbation, this contradicts to Lemma 2.3.1. Therefore, ϕ satisfies the conditions (a′) and (b′), this completes the proof.

In the same paper [Y1] K.Yano constructed an example of a dynamical system $\phi \in Z(S^1)$ which has the POTP but is not topologically stable. This result shows that the expansiveness condition in Theorem 2.1.3 is essential.

Let us describe the example of K.Yano. First consider a homeomorphism $\phi_0 : [0,1] \to [0,1]$ defined by

$$\phi_0(\alpha) = \begin{cases} \frac{1}{2}\alpha, & 0 \leq \alpha \leq \frac{1}{4}; \\ \frac{3}{2}\alpha - \frac{1}{4}, & \frac{1}{4} < \alpha < \frac{3}{4}; \\ \frac{1}{2}\alpha + \frac{1}{2}, & \frac{3}{4} \leq \alpha \leq 1. \end{cases}$$

For a positive number n, let

$$p_n = 2^{-n}, p_{-n} = 1-2^{-n}, q_n = 3\cdot 2^{-n-2}, q_{-n} = 1 - 3\cdot 2^{-n-2}.$$

Then the desired homeomorphism is given as follows:

$$\phi(0) = 0,$$

$$\phi(\alpha) = \begin{cases} p_{n+1} + 2^{-n-1}\phi_0[2^{n+1}(\alpha - p_{n+1})], & p_{n+1} \le \alpha \le p_n; \\ p_{-n} + 2^{-n-1}\phi_0[2^{n+1}(\alpha - p_{-n})], & p_{-n} \le p_{-n-1}; \end{cases}$$

$$n = 1, 2, \ldots, .$$

Fix n, and note that

$$\phi(p_n) = 2^{-n-1} + 2^{-n-1}\phi_0[2^{n+1}(\alpha - p_{n+1})] \mid_{\alpha=p_n} =$$

$$2^{-n-1}[1 + \phi_0(2^{n+1}(2^{-n} - 2^{-n-1}))] = 2^{-n} = p_n.$$

Hence, $\text{Fix}(\phi)$ is an infinite set. It follows from Theorem 2.3.1 that ϕ is not topologically stable.

So we have only to show that ϕ has the POTP.

Fix an arbitrary $\epsilon > 0$ and choose a positive integer n satisfying $2^{-n} < \epsilon$. Let $I = [p_{-n}, p_n]$, $J = [p_{n+1}, p_{-n-1}]$, and $J' = [p_{n+2}, p_{-n-2}]$. Then $J \subset J'$ and $I \cup J = S^1$. It follows from the definition of ϕ that there exists a homeomorphism ψ of S^1 which is topologically conjugate to some Morse-Smale diffeomorphism and satisfies $\psi \mid_{J'} = \phi \mid_{J'}$.

Theorem A.1 and Lemma 2.1.2 imply that ψ has the POTP. Hence we can find $\delta > 0$ satisfying the following two conditions:

(1) every δ-trajectory of ψ is 2^{-n-2}-traced by a trajectory of ψ;

(2) $\delta 2^{n+4} < 1$.

Take a δ-trajectory $\{x_k\}$ for ϕ. Consider two possible cases.

Case 1. $x_k \in I$ for any $k \in \mathbf{Z}$. In this case

$$d(x_k, \phi^k(0)) \le 2^{-n} < \epsilon, k \in \mathbf{Z}.$$

Case 2. There exists m such that $x_m \notin I$. We claim that in this case $x_k \in J$ for all $k \in \mathbf{Z}$. Let us show that if $x_0 \notin I$ then for $k \ge 0$ the sequence $\{x_k\}$ does not "jump" over the intervals $[p_n - 2^{-n-3}, p_n + 2^{-n-3}]$ and $[p_{-n} - 2^{-n-3}, p_{-n} + 2^{-n-3}]$ and hence $x_k \in J$. Suppose that for some $k \ge 0$ $x_k \ge p_n$, and $x_{k+1} < p_n$. It follows from the definition of ϕ that $\phi(x_k) \ge p_n$. As $d(x_{k+1}, \phi(x_k)) < 2^{-n-4}$ we obtain that

$$x_{k+1} \in (p_n - 2^{-n-3}, p_n).$$

Taking into account that $\phi'(\alpha) = \frac{1}{2}$ for $\alpha \in (p_n - 2^{-n-3}, p_n)$ we see that

$$d(\phi(x_{k+1}), p_n) = \frac{1}{2} d(x_{k+1}, p_n) \le 2^{-n-4}.$$

As $d(x_{k+2}, \phi(x_{k+1})) < 2^{-n-4}$ we obtain that either $x_{k+2} \ge p_n$ or $x_{k+2} \in (p_n - 2^{-n-3}, p_n)$, and so on. The case of the interval $[p_{-n} - 2^{-n-3}, p_{-n} + 2^{-n-3}]$ is considered similarly.

Analogous reasons show that if $x_0 \in I$ then for $k \le 0$ the sequence $\{x_k\}$ does not "jump" over the intervals $[q_n - 2^{-n-3}, q_n + 2^{-n-3}]$ and $[q_{-n} - 2^{-n-3}, q_{-n} + 2^{-n-3}]$.

Therefore the sequence $\{x_k\}$ is a δ-trajectory of ψ lying in J. Hence there exists a point $x \in S^1$ such that

$$d(x_k, \psi^k(x)) < 2^{-n-2}.$$

As $x_k \in J$ for $k \in \mathbf{Z}$, $\psi^k(x) \in J'$ for $k \in \mathbf{Z}$, so that $\psi^k(x) = \phi^k(x)$ for all k, and finally we see that

$$d(x_k, \phi^k(x)) < \epsilon, k \in \mathbf{Z}.$$

Thus we have shown that every δ-trajectory of ϕ is ϵ-traced by some trajectory of ϕ, hence ϕ has the POTP.

2.4 Density of Topologically Stable Systems in CLD(M)

This section is devoted to the following result first obtained by M.Shub in [Shu2].

Theorem 2.4.1. *Any $\phi \in$ Diff$^1(M)$ can be isotoped to a diffeomorphism ψ satisfying the STC by an isotopy which ia arbitrarily small in the C^0-topology.*

Let us explain that the existence of an isotopy of C^0-size ϵ between $\phi, \psi \in$ Diff$^1(M)$ means the following: there is an arc $\chi : [0,1] \to$ Diff$^1(M)$ such that $\chi(0) = \phi, \chi(1) = \psi$, and

$$\rho_0(\chi(t_1), \chi(t_2)) \leq \epsilon \text{ for } t_1, t_2 \in [0,1].$$

It was S.Smale who generalized the horseshoe construction to show that any isotopy class in Diff$^1(M)$ contains a diffeomorphism which is Ω-stable [Sm4]. After that R.Williams and M.Shub extended Smale's argument to replace Ω-stable by structurally stable, and M.Shub proved Theorem 2.4.1.

It follows immediately from Theorem 2.4.1 that diffeomorphisms which satisfy the STC are dense in CLD(M). Combining this result with Theorem 2.2.3 we obtain the following statement.

Theorem 2.4.2. *Topologically stable dynamical systems are dense in CLD(M).*

We don't give here a complete proof of Theorem 2.4.1. The reader is referred to the paper [Shu3] which contains a detailed proof of this result. In this book we explain how to construct an isotopy of small C^0-size between an arbitrary diffeomorphism ϕ and a diffeomorphism ψ which satisfies the STC in the case dim $M = 2$. The choice of dim $M = 2$ allows us to visualize main ideas of the construction and to avoid multidimensional topological difficulties (the correspondent techniques is described in detail in [Shu3]).

We begin with some simple facts from differential topology. We denote below by D^m an m-dimensional disc. As everywhere in this book, $S^{m-1} = \partial D^m$. Recall that a *handle decomposition*, H, of a manifold M with dim$M = n$, is a sequence of submanifolds with boundary

$$H : \emptyset \subset M_0 \subset \ldots \subset M_n = M, \tag{2.31}$$

where

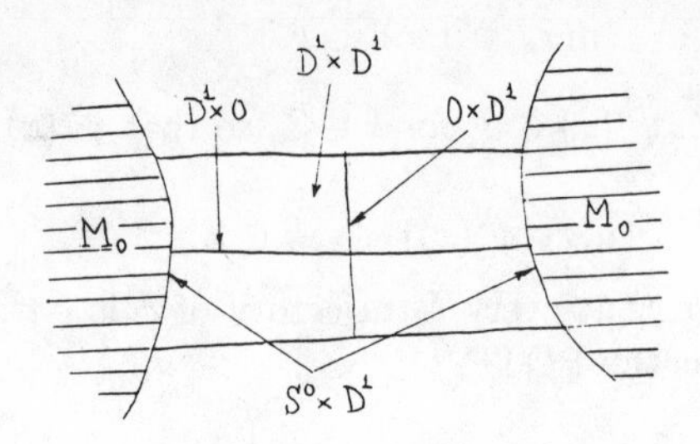

Figure 2.1

$$\overline{M_j \setminus M_{j-1}} = \bigcup_{i=1}^{m_j} D_i^j \times D_i^{n-j}, j = 1, \ldots, n,$$

and the $D_i^j \times D_i^{n-j}$ are attached to the boundary of M_{j-1} by disjoint embeddings of the $S_i^{j-1} \times D_i^{n-j}$.

Definition 2.7 *We say that $\phi \in Diff^1(M)$ preserves the decomposition (2.31) if*

$$\phi(M_j) \subset IntM_j, 0 \leq j \leq n.$$

Given a handle decomposition (2.31) we call the discs $D_i^j \times p, p \in D_i^{n-j}$, core discs, and the discs $q \times D_i^{n-j}, q \in D_i^j$, transverse discs.

In our case dim $M = 2$, a 0-handle $D^0 \times D^2$ is a 2-dimensional disc D^2 (D^0 is a point). Any point $p \in D^2$ is a core disc, and there is one transverse disc – D^2 itself.

We can visualize a 1-handle $D^1 \times D^1$ as follows (see Fig. 2.1). A 2-handle $D^2 \times D^0$ is also a disc D^2 being a core disc.

It is easy to understand that in our case dim$M = 2$ any core disc which intersects a core disc of lower dimension contains it (handle decompositions having this property are called fitted in [Shu3]).

Fix a diffeomorphism ϕ of M.

Definition 2.8 *Given a handle decomposition H, we say that ϕ is "fitted" with respect to H if for any core disc d, $\phi(d)$ contains any core disc it intersects.*

Let us now describe how to construct for a given diffeomorphism ϕ a handle decomposition H and an isotopy between ϕ and a diffeomorphism ψ such that ψ preserves H.

Given $\epsilon > 0$ for our manifold M we can find a smooth triangulation $T = \{T_i^j\}$ such that for any j-simplex T_i^j we have

$$\text{diam} T_i^j < \epsilon \tag{2.32}$$

(see [Mu2]). As usually we call

$$T^j = \bigcup_i T_i^j$$

the j-skeleton of T.

Fix such a triangulation

$$T = \{T_i^j : j \in \{0,1,2\}, i \in \{1,\dots,m_j\}\}$$

for M with small $\epsilon > 0$.

Fix small neighborhoods U_i^j of T_i^j in M, let

$$U^j = \bigcup_{i=1}^{m_j} U_i^j, j = 0,1,2.$$

It is geometrically evident that we can take U_i^j so that the sets U_i^0 are 0-handles of a handlebody decomposition H of M, the sets $\overline{U_i^1} \setminus U^0$ are 1-handles, and the sets $\overline{U_i^2} \setminus U^1$ are 2-handles.

Now for our diffeomorphism ϕ we can find points

$$r_i^2 \in \mathrm{Int}T_i^2, r_i^1 \in \mathrm{Int}T_i^1$$

(for any simplex T_i^j we take $\mathrm{Int}T_i^j$ in the inner topology) such that

$$r_i^2 \notin \phi(T^1), i = 1,\dots,m_1;$$

$$r_i^1 \notin \phi(T^0), i = 1,\dots,m_0.$$

We may fix U_i^j so small that

$$r_i^2 \notin \phi(\bar{U}^1), r_i^1 \notin \phi(\bar{U}^0).$$

By pushing away from points r_i^2 we can isotope ϕ to a diffeomorphism ψ_1 such that

$$\psi_1(\bar{U}^1) \subset U^1. \tag{2.33}$$

After that preserving (2.33) and pushing away from points r_i^1 we can isotope ψ_1 to a diffeomorphism ψ_2 such that

$$\psi_2(\bar{U}^2) \subset U^2,$$

hence ψ_2 preserves H. It is geometrically evident that the C^0-size of the isotopy between ϕ and ψ_2 depends only on ϵ in (2.32).

Let us show that we can find an isotopy between ψ_2 and a diffeomorphism ψ which is fitted with respect to H. It is easy to understand that we must pay attention only to one-dimensional core discs.

Consider a 1-handle $h_i = D_i^1 \times D_i^1$. As ψ_2 preserves H we have

$$\psi_2(h_i) \subset \mathrm{Int}M_1.$$

Fix in each 1-handle $h_i = D_i^1 \times D_i^1, i = 1,\dots,m_1$, a transverse disc, let it be $0 \times D_i^1$. It follows from the Transversality Theorem (see [Hi1]) that we can

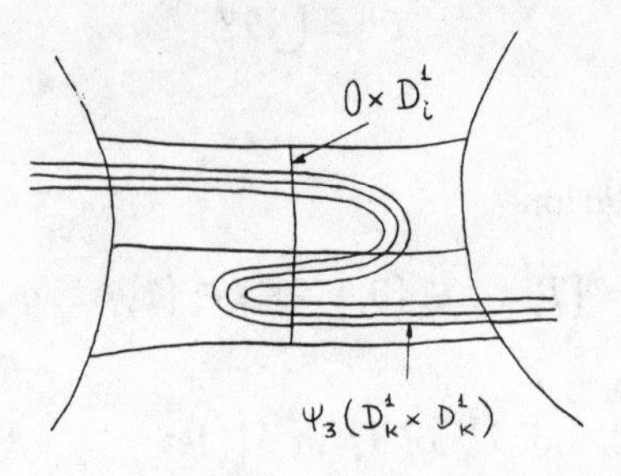

Figure 2.2

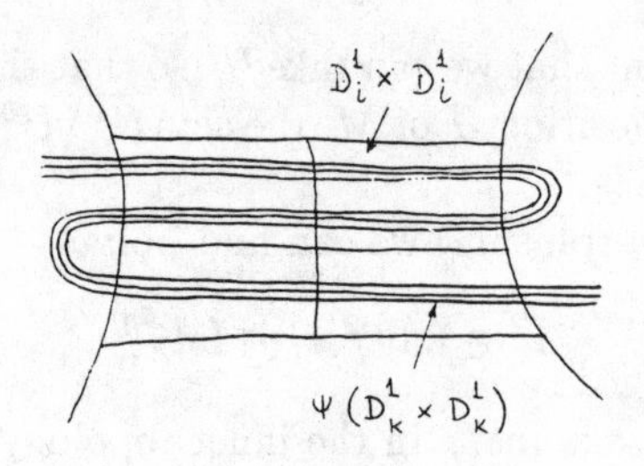

Figure 2.3

isotope ψ_2 to ψ_3 so that for any $i, k \in \{1, \dots, m_1\}$ $\psi_3(D_k^1 \times 0)$ is transversal to $0 \times D_i^1$. Now we can take neighborhoods U_k^1 of $T_k^1, k = 1, \dots, m_1$, so small that the analogue of (2.33) holds for ψ_3, and for any core disc $D_k^1 \times p$ in $D_k^1 \times D_k^1$, its image $\psi_3(D_k^1 \times p)$ is transversal to any $0 \times D_i^1$ (see Fig. 2.2).

Now we isotope ψ_3 to ψ by pushing away from the transverse discs $0 \times D_i^1$ along core discs $D_i^1 \times p$ to obtain ψ being fitted with respect to H (compare Fig. 2.2 and Fig. 2.3).

In our isotopy of ϕ into ψ we obviously can make ψ uniformly expanding in core discs and uniformly contracting in transverse discs.

If we define sets

$$\Lambda_j(\psi) = \bigcap_{k \in \mathbf{Z}} \psi^k(\overline{M_j \setminus M_{j-1}}), j = 0, 1, 2,$$

then it follows from our construstion that these sets have hyperbolic structure. The same reasons as in the proof of Lemma 1.4.1 show that if we set

$$\Omega_j(\psi) = \Omega(\psi) \cap (M_j \setminus M_{j-1}), j = 0, 1, 2,$$

then

$$\Omega_j(\psi) \subset \Lambda_j(\psi), j = 0, 1, 2.$$

Hence the set $\Omega(\psi)$ is hyperbolic. For points of $\Lambda_j(\psi)$ the local unstable manifolds are the core discs, and the local stable manifolds are transverse discs through these points. It is evident now that ψ satisfies the geometric strong transversality condition. It is well-known (see [Ni1]) that the hyperbolicity of $\Omega(\psi)$ and the geometric strong transversality condition imply the density of $\mathrm{Per}(\psi)$ in $\Omega(\psi)$. So, ψ satisfies Axiom A. It follows now from Theorem 0.4.8 that ϕ satisfies the STC.

Let us now prove Theorem 1.1.3. Fix arbitrary $\epsilon > 0$ and define the subset V_ϵ of CLD(M) as follows: $\phi \in V_\epsilon$ if there is a neighborhood W of ϕ in $Z(M)$ such that for any two systems $\phi_1, \phi_2 \in W$ we have

$$\bar{R}(\Theta(\phi_1), \Theta(\phi_2)) < \epsilon$$

(the map Θ was defined in Sect. 1.1). It follows from the definition that any $\psi \in W \cap \mathrm{CLD}(M)$ belongs to V_ϵ, hence V_ϵ is open in $\mathrm{CLD}(M)$ for any $\epsilon > 0$.

If ϕ is a topologically stable diffeomorphism then it follows from Theorem 2.1.1 that ϕ is a point of continuity for Θ, hence $\phi \in V_\epsilon$ for any $\epsilon > 0$. By Theorem 2.4.2 the set V_ϵ is dense in CLD(M).

It remains to note that the set

$$V = \bigcap_{\epsilon>0} V_\epsilon$$

is a residual subset of CLD(M), and that any system $\phi \in V$ is evidently tolerance-$Z(M)$-stable. This completes the proof of Theorem 1.1.3.

2.5 Two Conditions of Topological Stability

We describe here without proofs two results on topologically stable systems.

M.Hurley studied in [Hu2] properties of the chain-recurrent set $\mathrm{CR}(\phi)$ of a topologically stable diffeomorphism ϕ (see Sect. 1.4 for definitions).

Let us introduce the following equivalence relation E on M : xEy if $x \in \mathrm{CH}(y, \phi)$, $y \in \mathrm{CH}(x, \phi)$. Classes of equivalence generated by E are called chain components of ϕ .

For a periodic point p of period m we define the set

$$W^s(p) = \{x \in M : d(\phi^{mk}(x), p) \to 0 \text{ as } k \to +\infty\}$$

being an analogue of the stable manifold for a hyperbolic periodic point.

Theorem 2.5.1 [Hu2]. *If ϕ is a topologically stable diffeomorphism then*

(a) $CR(\phi) = Per(\phi)$;

(b) there is only finitely many chain components each having a dense orbit;

(c) there is a finite set of periodic points $p_1, \ldots, p_m$ such that

$$M = \overline{W^s(p_1)} \cup \ldots \cup \overline{W^s(p_m)};$$

(d) if X is a chain component of ϕ, and m is the least period of periodic points in X, then X contains not more than m chain components of $g = \phi^m$;

in this case for any chain component T of g the following holds – if U, V are non-empty open sets in T then there is $k_0 > 0$ such that

$$g^k(U) \cap V \neq \emptyset$$

for $k > k_0$.

This result shows that topologically stable diffeomorphisms have properties analogous to well-known properties of structurally stable diffeomorphisms (see Theorems 0.4.2, 0.4.3, 0.4.4).

J.Lewowicz applied in [L] ideas of Lyapunov-type functions to give suffucient conditions for the topological stability of a diffeomorphism ϕ of class C^1.

To formulate the theorem of J.Lewowicz we need a construction from [Pa4] that enables us to relate diffeomorphisms on the manifold M with vector fields on a manifold $\tilde{M}$ of dimension one higher. This construction is called a suspension of a diffeomorphism.

Let $\tilde{M}$ be a smooth manifold and let Σ be a compact submanifold of codimension 1.

Take a vector field X of class C^1 on $\tilde{M}$ and let $\Phi : \mathbf{R} \times \tilde{M} \to \tilde{M}$ be the flow generated by X.

Definition 2.9 *We say that Σ is a "global transversal section" for X if*

(a) X is transversal to Σ;

(b) for any point $x \in \Sigma$ the positive trajectory $\{\Phi(t, x) : t > 0\}$ returns to intersect Σ again.

If Σ is a global transversal section for a vector field X then Φ induces a diffeomorphism $f : \Sigma \to \Sigma$ which associates to each point $p \in \Sigma$ the point $f(p)$ where the positive trajectory of p first intersects Σ. The diffeomorphism f is called the Poincaré map associated with Σ.

The following statement is obtained in [Pa3].

Lemma 2.5.1. *Let ϕ be a diffeomorphism of class C^1 on M. Then we can find a manifold $\tilde{M}$, a vector field X of class C^1 on $\tilde{M}$ admitting a global transversal section Σ, and a diffeomorphism $h : M \to \Sigma$ of class C^1 that conjugates ϕ and the Poincaré map $f : \Sigma \to \Sigma$.*

Any vector field X which has the properties described by Lemma 2.5.1 is called a suspension of ϕ. Let for $t \in \mathbf{R}$

$$\Sigma_t = \Phi(t, \Sigma).$$

It can be shown that

$$\tilde{M} = \bigcup_{t \in \mathbf{R}} \Sigma_t$$

(see [Pa4]). Fix a Riemannian metric $\tilde{d}$ on $\tilde{M}$.

Theorem 2.5.2 [L]. *Let ϕ be a diffeomorphism of class C^1 on M, and let X be a suspension of ϕ of class C^1 on $\tilde{M}$.*

Assume that there is a number $\alpha > 0$ and a real continuous function V defined on the set

$$\{(x,y) \in \tilde{M} \times \tilde{M} : x, y \in \Sigma_t \text{ for some } t \in \mathbf{R}, \text{ and } \tilde{d}(x,y) < \alpha\}$$

such that:

(a) $V(x,x) = 0, V(x,y) > 0$ if $x \neq y$;

(b) the functions

$$\dot{V}(x,y) = \lim_{t \to 0} \frac{1}{t}(V(\Phi(t,x), \Phi(t,y)) - V(x,y)),$$

$$\ddot{V}(x,y) = \lim_{t \to 0} \frac{1}{t}(\dot{V}(\Phi(t,x), \Phi(t,y)) - \dot{V}(x,y))$$

are continuous;

(c) $\ddot{V}(x,y) > 0$ if $x \neq y$;

(d) there is $\rho > 0$ which has the following property:

for $x \in \tilde{M}$ (with $x \in \Sigma_t, t \in \mathbf{R}$) there are subspaces S_x, U_x of $T_x\Sigma_t$ such that:

(d.1) $S_x \oplus U_x = T_x\Sigma_t$;

(d.2) S_x, U_x depend continuously on x;

(d.3) if $v \in S_x, 0 < |v| \leq \rho$ then

$$\dot{V}(x, exp_x v) < 0;$$

(d.4) if $v \in U_x, 0 < |v| \leq \rho$ then

$$\dot{V}(x, exp_x v) > 0;$$

(e) either for each $x \in \Sigma_t$ the subspace S_x intersects trivially the tangent space at x of any manifold contained in

$$\{y \in \Sigma_t : \dot{V}(x,y) \geq 0\},$$

or for each $x \in \Sigma_t$ the subspace U_x intersects trivially the tangent space at x of any manifold contained in

$$\{y \in \Sigma_t : \dot{V}(x,y) \leq 0\}.$$

Then ϕ is topologically stable.

It is shown in [L] that Anosov diffeomorphisms (that is diffeomorphisms ϕ such that M is a hyperbolic set) satisfy the assumptions of Theorem 2.5.1. J.Lewowicz gives in [L] interesting examples of diffeomorphisms which satisfy the assumptions of Theorem 2.5.1 and fail to be Anosov.

3. Perturbations of Attractors

3.1 General Properties of Attractors

Asymptotically stable compact invariant sets (attractors) for dynamical systems were studied by many authors (see [Bh, Con, Hu1, Mi, Pl1, Sha]).

We are mostly interested here in various types of stability of attractors with respect to C^0-small perturbations of the system. Let us begin with exact definitions.

Definition 3.1 *We say that a subset $I \subset M$ is "Lyapunov stable" with respect to a dynamical system $\phi \in Z(M)$ if*

(a) I is compact and ϕ-invariant;

(b) given any neighborhood U of I there exists a neighborhood V of I such that

$$\phi^k(V) \subset U \text{ for } k \geq 0.$$

Definition 3.2 *We say that a subset $I \subset M$ is an "attractor" of a system ϕ if*

(a) I is Lyapunov stable with respect to ϕ;

(b) there exists a neighborhood W of I such that for any $x \in W$

$$\lim_{k\to\infty} d(\phi^k(x), I) = 0. \tag{3.1}$$

Note that it follows from our definitions that the manifold M is an attractor for any $\phi \in Z(M)$.

Various authors give different definitions of an attractor, so the reader must always pay attention to exact definitions reading papers devoted to attractors. Often definitions of an attractor include such properties as indecomposability or topological transitivity. It is useful to read the paper [Mi] containing a discussion of possible definitions and their relations to applications.

For an attractor I we denote by J its boundary and by $D(I)$ its *basin of attraction* (sometimes we say simply "basin of I"), that is the set

$$D(I) = \{x \in M : \lim_{k\to\infty} d(\phi^k(x), I) = 0\}.$$

We establish here some simple properties of attractors we need.

Lemma 3.1.1. *$D(I)$ is open.*

Proof. Fix $x \in D(I)$. As (3.1) holds we can find $m > 0$ such that $\phi^m(x) \in W$ (here W is a neighborhood of I described in the definition of an attractor). Then there exists a neighborhood N of x such that $\phi^m(N) \subset W$. Evidently $N \subset D(I)$.

Lemma 3.1.2. *Let K be a compact subset of $D(I)$. Then given a neighborhood N of I there exists m_0 such that $\phi^m(K) \subset N$ for $m \geq m_0$.*

Proof. To obtain a contradiction suppose that there exists a neighborhood N of I such that for any m_0

$$\phi^m(K) \setminus N \neq \emptyset \tag{3.2}$$

for some $m \geq m_0$. Find a sequence of points $x_m \in K$ such that $\phi^m(x_m) \notin N$. Let x_0 be a limit point of the sequence x_m. For simplicity we consider the case

$$\lim_{m \to \infty} x_m = x_0.$$

Then $x_0 \in K$, hence $x_0 \in D(I)$. As the set I is Lyapunov stable there is a neighborhood V of I such that $\phi^k(V) \subset N$ for $k \geq 0$.

Find k_0 such that $\phi^{k_0}(x_0) \in V$. Then for large m we have $\phi^{k_0}(x_m) \in V$, hence

$$\phi^{k+k_0}(x_m) \in N \text{ for } k \geq 0.$$

Takihg $m > k_0$ we obtain a contradiction with (3.2). This completes the proof.

Corollary *If I is an attractor, $x \notin I$, and K is a compact subset of $D(I)$ then there exists m_0 such that $\phi^m(x) \notin K$ for $m \leq m_0$.*

Proof. As $x \notin I$ and I is compact there is a neighborhood N of I such that $x \notin N$. Apply Lemma 3.1.2 and find m_0 such that

$$\phi^m(K) \subset N \text{ for } m \geq m_0.$$

Now we describe a possible way to construct attractors.

Lemma 3.1.3. *Let U be an open subset of M . Assume that for a dynamical system ϕ there is a natural N such that*

$$\phi^N(\overline{U}) \subset U, \phi^{N+1}(\overline{U}) \subset U. \tag{3.3}$$

Then the set

$$I = \bigcap_{k \geq 0} \phi^{kN}(\overline{U}) \tag{3.4}$$

is an attractor for ϕ , and $\overline{U} \subset D(I)$.

Proof. It follows from (3.3) that

$$\overline{U} \supset \phi^N(\overline{U}) \supset \phi^{2N}(\overline{U}) \supset \dots \tag{3.5}$$

hence I is non-empty and compact. Take natural l and note that

$$(l+1)(N+1)N = l(N+1)N + N^2 + N.$$

Hence

$$\phi^{(l+1)(N+1)N}(\overline{U}) = \phi^{l(N+1)N+N^2}(\phi^N(\overline{U})) \subset$$
$$\subset \phi^{l(N+1)N+N^2}(\overline{U}).$$

Similarly we show that

$$\phi^{(l+1)(N+1)N}(\overline{U}) \subset \phi^{l(N+1)N+N(N-1)}(\overline{U}) \subset \dots \subset \phi^{l(N+1)N}(\overline{U}). \tag{3.6}$$

Using (3.5),(3.6) we obtain that

$$I = \bigcap_{l>0} \phi^{l(N+1)N}(\overline{U}).$$

Apply the second inclusion in (3.3) to define the compact

$$I_1 = \bigcap_{k\geq 0} \phi^{k(N+1)}(\overline{U}).$$

Analogous considerations show that

$$I_1 = \bigcap_{l>0} \phi^{l(N+1)N}(\overline{U}),$$

hence $I = I_1$. Now note that

$$\phi^N(I) = \bigcap_{k\geq 1} \phi^{kN}(\overline{U}) = I,$$

similarly

$$\phi^{N+1}(I) = \phi^{N+1}(I_1) = I_1 = I,$$

hence

$$\phi^{N+1}(I) = \phi^N(I) \text{ , and } \phi(I) = I.$$

So, I is ϕ-invariant.

Let us now show that I is Lyapunov stable. Fix a neighborhood W of I . As I is invariant and compact there is a neighborhood V_0 of I such that

$$\phi^m(V_0) \subset W \text{ for } m = 0, 1, \dots, N-1.$$

It follows from (3.4),(3.5) that there is m_0 such that for $m \geq m_0$

$$\phi^{mN}(U) \subset V_0.$$

Set

$$V = \phi^{m_0 N}(U).$$

Take now arbitrary $k \geq 0$. We can write

$$k = lN + l_1, \text{ where } l \geq 0, 0 \leq l_1 \leq N-1.$$

Then

$$\phi^k(V) = \phi^{l_1}(\phi^{lN}(V)) =$$
$$\phi^{l_1}(\phi^{(m_0+l)N}(U)) \subset \phi^{l_1}(V_0) \subset W.$$

So, I is Lyapunov stable. The property (b) in the definition of an attractor is established analogously. The inclusion $D(I) \supset \overline{U}$ is evident.

Remark. If instead of (3.3) we assume that

$$\phi(\overline{U}) \subset U$$

then the same proof shows that the set

$$I = \bigcap_{k \geq 0} \phi^k(\overline{U})$$

is an attractor for ϕ with $\overline{U} \subset D(I)$.

Definition 3.3 *We say that a non-empty compact set I is a "quasi-attractor" for a system $\phi \in Z(M)$ if there exists a countable family $\{I_m : m \geq 0\}$ of attractors of ϕ such that*

$$I = \bigcap_{m \geq 0} I_m. \tag{3.7}$$

We define the basin $D(I)$ for a quasi-attractor I using (3.1). Note that for a quasi-attractor its basin is not always open. Consider the following example. Let $M = S^1$ with coordinate $\alpha \in [0,1)$. Consider a scalar function $F \in C^\infty(S^1)$ such that

$$F(\alpha) = 0 \text{ for } \alpha \in \{0, 1/2, 1/3, \ldots\},$$
$$F(\alpha) > 0 \text{ for } \alpha \in (1/2, 1),$$
$$F(\alpha) < 0 \text{ for } \alpha \in (1/3, 1/2) \cup (1/4, 1/3) \cup \ldots.$$

Let $f(t, \alpha)$ be the flow generated by the system of differential equations

$$\frac{d\alpha}{dt} = F(\alpha),$$

and let $\phi(\alpha) = f(1, \alpha)$.

Then evidently for $m = 0, 1, \ldots$ the set

$$I_m = [0, \frac{1}{m+3}]$$

is an attractor of ϕ, and

$$I = \bigcap_{m \geq 0} I_m = \{0\}$$

is a quasi-attractor. Its basin $D(I)$ is the set $(1/2, 1) \cup \{0\}$, so $D(I)$ is not open.

It is easy to see that if we let by definition the empty set to be an attractor then for any finite family of attaractors their intersection is again an attractor.

Let $\{I_m, m \geq 0\}$ be a family of attractors, and let I given by (3.7) be a quasi-attractor. Introduce for $k \geq 0$

$$I_k' = \bigcap_{m=0}^{k} I_m.$$

Then evidently

$$I_0' \supset \ldots \supset I_k' \supset I_{k+1}' \supset \ldots,$$

and

$$I = \bigcap_{k \geq 0} I_k'.$$

Below for any quasi-attractor I defined by (3.7) we consider families $\{I_m\}$ with $I_m \supset I_{m+1}, m \geq 0$.

3.2 Stability of Attractors for Generic Dynamical Systems

Let S be a metric on M^*.

Definition 3.4 *We say that an attractor I of a dynamical system ϕ is "stable in $Z(M)$ with respect to S" if given $\epsilon > 0$ there exists a neighborhood W of ϕ in $Z(M)$ such that any system $\tilde{\phi}$ in W has an attractor $\tilde{I}$ with*

$$S(I, \tilde{I}) < \epsilon.$$

M.Hurley in [Hu1] studied stability of attractors in $Z(M)$ with respect to the Hausdorff metric R on M^* (note that in [Hu1] this property of attractors was called persistence). He proved the following result.

Theorem 3.2.1 [Hu1]. *A generic system $\phi \in Z(M)$ has the property : any attractor I of ϕ is stable in $Z(M)$ with respect to R .*

Stability of attractors with respect to R_0 was investigated in [Pi3]. Let us prove the following theorem which is a generalization of Theorem 3.2.1.

Theorem 3.2.2 [Pi3]. *A generic system $\phi \in Z(M)$ has the property : any attractor I of ϕ is stable in $Z(M)$ with respect to R_0.*

We begin by proving a simple

Lemma 3.2.1. *Let $I \in M^*$. Then*

$$\lim_{\delta \to 0} R(M \setminus N_\delta(I), \overline{M \setminus I}) = 0.$$

Proof. As evidently

$$M \setminus N_\delta(I) \subset \overline{M \setminus I}$$

we have only to show that

$$\lim_{\delta \to 0} r(\overline{M \setminus I}, M \setminus N_\delta(I)) = 0. \tag{3.8}$$

To get a contradiction suppose that for $k > 0$ there exists $a > 0$ and points $\xi_k \in \overline{M \setminus I}$ such that

$$d(\xi_k, M \setminus N_{1/k}(I)) \geq a.$$

Let ξ be a limit point of the sequence ξ_k , then for large k we have

$$d(\xi, M \setminus N_{1/k}(I)) \geq \frac{a}{2}. \tag{3.9}$$

If $\xi \notin I$ then $\xi \notin N_{1/k}(I)$ for large k , so that (3.9) is impossible. If $\xi \in I$ then evidently $\xi \in \partial I$. Find a sequence $x_l \in M \setminus I$ such that $\lim_{l \to \infty} x_l = \xi$, then for some l_0 we have

$$d(x_{l_0}, \xi) < \frac{a}{2}. \tag{3.10}$$

As $x_{l_0} \notin I$ we obtain that $x_{l_0} \notin N_{1/k}(I)$ for large k , so we obtain a contradiction between (3.9) and (3.10) . This proves (3.8) .

To prove Theorem 3.2.2 consider a non-empty open set $U \subset M$ and define two maps

$$G^+, G^- : Z(M) \to M^*$$

in the following way. If a system $\phi \in Z(M)$ has an attractor I such that

$$I \subset U \subset \overline{U} \subset D(I) \tag{3.11}$$

let

$$G^+(\phi) = I, G^-(\phi) = \overline{M \setminus I}.$$

If ϕ has no attractors satisfying (3.11) let $G^+(\phi) = M, G^-(\phi) = \emptyset$.

To show that the maps G^+, G^- are properly defined let us show that any system has not more than one attractor satisfying (3.11). To get a contradiction suppose that $\phi \in Z(M)$ has two attractors I_1, I_2 such that

$$I_1 \subset U \subset \overline{U} \subset D(I_1),$$

$$I_2 \subset U \subset \overline{U} \subset D(I_2),$$

and $I_1 \neq I_2$. Let $I_1 \setminus I_2 \neq \emptyset$, and fix a point $x \in I_1 \setminus I_2$. As I_1 is invariant, $O(x,\phi) \subset I_1 \subset \overline{U}$. Apply corollary of Lemma 3.1.2 to find $m \in \mathbf{Z}$ with $\phi^m(x) \notin \overline{U}$. The contradiction we obtained shows that G^+, G^- are properly defined.

Let us now show that the map G^+ is upper semi-continuous, and the map G^- is lower semi-continuous.

We begin considering G^+ . Fix $\phi \in Z(M)$. If $G^+(\phi) = M$ then the upper semi-continuity of G^+ at ϕ is trivial. Let (3.11) is satisfied, and $G^+(\phi) = I$. Take arbitrary $\epsilon > 0$ (we consider ϵ so small that $N_\epsilon(I) \subset U$). It follows from Lemma 3.1.2 that we can find a natural N with

$$\phi^N(\overline{U}) \subset N_\epsilon(I), \phi^{N+1}(\overline{U}) \subset N_\epsilon(I).$$

There exists a neighborhood W of ϕ in $Z(M)$ such that for any $\tilde{\phi} \in W$

$$\tilde{\phi}^N(\overline{U}) \subset N_\epsilon(I), \tilde{\phi}^{N+1}(\overline{U}) \subset N_\epsilon(I).$$

Apply Lemma 3.1.3 to find an attractor $\tilde{I}$ of $\tilde{\phi}$ such that

$$\tilde{I} \subset N_\epsilon(I) \subset U \subset \overline{U} \subset D(\tilde{I}).$$

Evidently $G^+(\tilde{\phi}) = \tilde{I}$, so G^+ is upper semi-continuous at ϕ .

Now consider the map G^- . If for $\phi \in Z(M)$ we have $G^-(\phi) = \emptyset$ then the lower semi-continuity of G_- at ϕ is trivial. Let for ϕ (3.11) is satisfied, and $G^-(\phi) = \overline{M \setminus I}$. Fix $\epsilon > 0$. Using Lemma 3.2.1 find $\delta > 0$ such that $N_\delta(I) \subset U$, and

$$R(M \setminus N_\delta(I), \overline{M \setminus I}) < \epsilon. \tag{3.12}$$

Reasons similar to the case of G^+ show that there exists a neighborhood W of ϕ in $Z(M)$ such that any system $\tilde{\phi} \in W$ has an attractor $\tilde{I}$ with

$$\tilde{I} \subset N_\delta(I) \subset U \subset \overline{U} \subset D(\tilde{I}).$$

We claim that in this case

$$\overline{M \setminus I} \subset N_\epsilon(\overline{M \setminus \tilde{I}}), \tag{3.13}$$

hence $G_-(\tilde{\phi}) \subset N_\epsilon(G^-(\phi))$.

Take $x \in \overline{M \setminus I}$. As $\tilde{I} \subset N_\delta(I)$ we have

$$M \setminus N_\delta(I) \subset M \setminus \tilde{I} \subset \overline{M \setminus \tilde{I}},$$

hence

$$d(x, \overline{M \setminus \tilde{I}}) \leq d(x, M \setminus N_\delta(I)) < \epsilon$$

(we take (3.12) into account here). This establishes (3.13), and the lower semi-continuity of G^- .

As the manifold M has a countable basis for its topology, we can find a countable family $\{U_m : m \in \mathbf{Z}\}$ of open sets in M having the following property :

for any two compact sets $K_1, K_2 \in M^*$ with $K_1 \cap K_2 = \emptyset$ there exists U_{m_0} such that

$$K_1 \subset U_{m_0}, K_2 \cap \overline{U}_{m_0} = \emptyset.$$

Define for any $U_m, m \in \mathbf{Z}$, corresponding maps G^+_m, G^-_m, and let $C^+_m, C^-_m \subset Z(M)$ be the sets of continuity points of G^+_m, G^-_m, respectively.

It follows from Lemma 0.2.1 that C^+_m, C^-_m are residual subsets of $Z(M)$. Then

$$C^* = \bigcap_{m \in \mathbf{Z}} (C^+_m \cap C^-_m)$$

is also a residual subset of $Z(M)$. We claim that for $\phi \in C^*$ any attractor is stable in $Z(M)$ with respect to R_0.

Take $\phi \in C^*$, and let I be an attractor of ϕ. The sets I and $M \setminus D(I)$ are disjoint and compact, find $U_{m_0} \in \{U_m\}$ such that

$$I \subset U_{m_0} \subset \overline{U}_{m_0} \subset D(I).$$

Fix $\epsilon > 0$. As $G^+_{m_0}, G^-_{m_0}$ are continuous at ϕ there exists a neighborhood W of ϕ in $Z(M)$ such that for any system $\tilde{\phi} \in W$

$$R(G^+_{m_0}(\tilde{\phi}), G^+_{m_0}(\phi)) < \epsilon, R(G^-_{m_0}(\tilde{\phi}), G^-_{m_0}(\phi)) < \epsilon. \tag{3.14}$$

It follows from the definitions of G^+, G^- that in this case $\tilde{I} = G^+_{m_0}(\tilde{\phi})$ is an attractor of $\tilde{\phi}$, and $G^-_{m_0} = \overline{M \setminus \tilde{I}}$. Hence, inequalities (3.14) imply

$$R_0(I, \tilde{I}) < \epsilon.$$

This completes the proof.

Remark. Fix again a countable family $\{U_m\}$ of open sets described in the proof. It was shown during the proof that for any $\phi \in Z(M)$:

(a) if I is an attractor then there exists U_m such that (3.11) (with $U = U_m$) holds;

(b) for any U_m there is not more than one attractor I such that (3.11) (with $U = U_m$) holds.

Hence, the set of attractors of ϕ is at most countable.

Attractors which are stable in $Z(M)$ with respect to R_0 have some important qualitative properties.

Theorem 3.2.3 [Pi3]. *If an attractor I of a system ϕ is stable in $Z(M)$ with respect to R_0 then its boundary J is Lyapunov stable.*

Proof. To get a contradiction suppose that J is not Lyapunov stable. In this case there exists $a > 0$ and sequences $x_m \in \text{Int} I$, t_m such that

$$\lim_{m \to \infty} t_m = +\infty, \lim_{m \to \infty} d(x_m, J) = 0, d(\phi^{t_m}(x_m), J) \geq a.$$

Passing to a subsequence if necessary find a point $p \in J$ such that $\lim_{m\to\infty} x_m = p$. Fix a sequence of points $y_m \in M \setminus I$ with $\lim_{m\to\infty} y_m = p$. Find $b > 0$ such that

$$\overline{N_b(I)} \subset D(I).$$

It follows from corollary of Lemma 3.1.2 that there exist numbers $\tau_m < 0$ having the following property :

$$\phi^{\tau_m}(y_m) \notin N_b(I).$$

Let

$$\tilde{x_m} = \phi^{t_m}(x_m), \tilde{y_m} = \phi^{\tau_m}(y_m).$$

Apply Lemma 0.3.3 to find a sequence of systems $\phi_m \in Z(M)$ and a sequence of numbers $\theta_m > 0$ such that

$$\lim_{m\to\infty} \rho_0(\phi_m, \phi) = 0, \phi_m^{\theta_m}(\tilde{x_m}) = \tilde{y_m}.$$

We assumed that I is stable with respect to R_0 hence for large m there exist attractors I_m of systems ϕ_m such that

$$\lim_{m\to\infty} R_0(I, I_m) = 0. \tag{3.15}$$

If $\tilde{x_m} \in I_m$ then

$$r(I_m, I) \geq b. \tag{3.16}$$

If $\tilde{x_m} \notin I_m$ then $\tilde{y_m} \notin I_m$ and consequently

$$r(M \setminus I_m, M \setminus I) \geq a. \tag{3.17}$$

Inequalities (3.16),(3.17) contradict to (3.15). This completes the proof.

Remark. [1] Let us give an example of an attractor I which is stable in $Z(M)$ with respect to R_0 and such that its boundary J is not an attractor. Let $M = S^1$ with coordinate $\alpha \in [0, 1)$. Consider a scalar function $F \in C^\infty(S^1)$ such that

$$F(\alpha) = 0 \text{ for } \alpha \in \{\frac{3}{4}, \frac{1}{2}, \ldots, 0\};$$

$$F(\alpha) > 0 \text{ for } \alpha \in (\frac{3}{4}, 1) \cup \bigcup_{m\geq 1} (\frac{1}{2m+1}, \frac{1}{2m});$$

$$F(\alpha) < 0 \text{ for } \alpha \in (\frac{1}{2}, \frac{3}{4}) \cup \bigcup_{m\geq 2} (\frac{1}{2m}, \frac{1}{2m-1}).$$

Let $f(t, \alpha)$ be the flow generated by the system of differential equations

$$\frac{d\alpha}{dt} = F(\alpha),$$

and let $\phi(\alpha) = f(1, \alpha)$. It is easy to see that $I = [0, 1/2]$ is an attractor of ϕ being stable in $Z(S^1)$ with respect to R_0 , but its boundary $J = \{0\} \cup \{1/2\}$ is not an attractor.

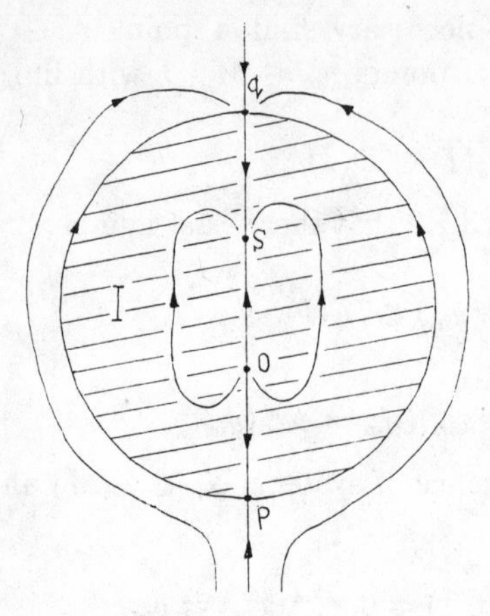

Figure 3.1

Remark. [2] It follows from Theorem 3.2.3 that the stability of an attractor with respect to R and the stability of an attractor with respect to R_0 are qualitatively different properties. To show this let us give an example of an attractor I being stable with respect to R and such that the set $J = \partial I$ is not Lyapunov stable.

Consider a coordinate neighborhood U in S^2 being homeomorphic to $\mathbf{R}^2$. Let Φ be a flow on S^2 which has the picture of trajectories in U shown in Fig. 3.1. Here o is a source, s is a sink, p is a saddle, and q is a degenerate rest point. The attractor I we are interested in is shaded. Given $\epsilon > 0$ we can construct closed lines L_1, L_2 (dotted in Fig. 3.2) so that if I_1, I_2 are the domains bounded by L_1, L_2 then

$$R(I_j, I) < \epsilon, j = 1, 2,$$

and L_1, L_2 have no contact with the trajectories of Φ . It is easy to understand that any flow $\tilde{\Phi}$ which is C^0-close to Φ has an attractor $\tilde{I}$ such that

$$I_2 \subset \tilde{I} \subset I_1,$$

hence $R(I, \tilde{I}) < \epsilon$. So, I is stable with respect to R, and J is not Lyapunov stable. To obtain the corresponding example for discrete dynamical systems take the time-one map of Φ.

The following result is a corollary of Theorems 3.2.2, 3.2.3.

Theorem 3.2.4 [Pi3]. *A generic system $\phi \in Z(M)$ has the property : for any attractor I of ϕ its boundary J is Lyapunov stable.*

M.Hurley studied in [Hu1] stability with respect to R not only for attractors but also for such objects as basins of attractors, quasi-attractors, chain transitive attractors, and chain transitive quasi-attractors.

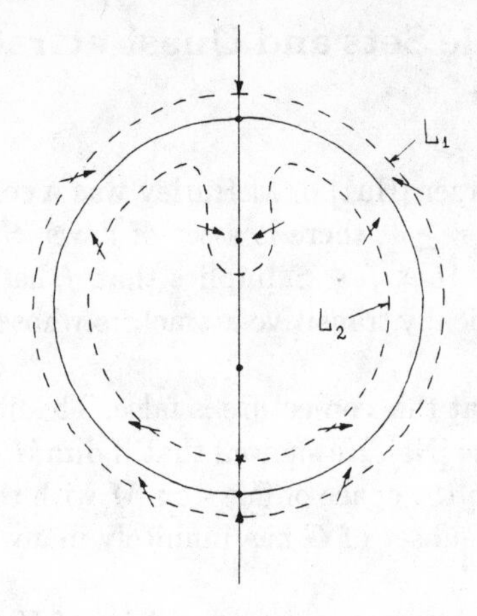

Figure 3.2

Definition 3.5 *We say that an attractor (or a quasi-attractor) I of a system $\phi \in Z(M)$ is "chain transitive" if $I \subset CR(\phi)$.*

Let S be a metric on M^*. Now we can define stability in $Z(M)$ with respect to S for a quasi-attractor (for a chain transitive attractor, or for a chain transitive quasi-attractor) I of a system $\phi \in Z(M)$ repeating the definition of stability of an attractor in $Z(M)$ with respect to S and putting everywhere in this definition words quasi-attractor (respectively, chain transitive attractor, chain transitive quasi-attractor) instead of the word attractor.

Theorem 3.2.5 [Hu1]. *A generic system $\phi \in Z(M)$ has the following property : any quasi-attractor (chain transitive attractor, chain transitive quasi-attractor) I of ϕ is stable in $Z(M)$ with respect to R .*

Theorem 3.2.6 [Hu1]. *A generic system $\phi \in Z(M)$ has the following property : if I is an attractor (a quasi-attractor, or a chain transitive attractor) of ϕ then given $\epsilon > 0$ there exists a neighborhood W of ϕ such that any system $\tilde{\phi} \in W$ has an attractor (respectively, a quasi-attractor, or a chain transitive attractor) $\tilde{I}$ with*

$$R(\overline{D(I)}, \overline{D(\tilde{I})}) < \epsilon.$$

We do not give proofs of Theorems 3.2.5,3.2.6 here, and refer the reader to the original paper [Hu1].

M.Moreva obtained in [More] analogues of Theorems 3.2.5, 3.2.6 for metric R_0 on M^* instead of R.

3.3 Lyapunov Stable Sets and Quasi-attractors in Generic Dynamical Systems

One of the origins of the paper [Hu1] of M.Hurley was a conjecture formulated by R.Thom in [Th] : given $r \geq 1$, there is a set of flows, S, residual in the set of all C^r flows on M, such that $f \in S$ implies that f has a finite number of structurally stable, topologically transitive attractors whose basins are dense in M.

It is well-known now that this conjecture is false. The first of the counterexamples is due to S.Newhouse [Ne]. He showed that if $\dim M \geq 3$, and $r \geq 2$ then there is an open subset G in the space of flows on M with the C^r-topology such that any flow in a residual subset of G has infinitely many hyperbolic periodic sinks.

Nevertheless, the following statement proved by M.Hurley in [Hu] can be considered as a weak variant of the conjecture of R.Thom.

Theorem 3.3.1. *If $\dim M \leq 3$ then a generic system $\phi \in Z(M)$ has the properties :*

(a) any chain transitive quasi-attractor of ϕ is stable in $Z(M)$ with respect to R ;

(b) the union of basins of chain transitive quasi-attractors of ϕ is a residual subset of M.

Proof. The first statement of the theorem follows from Theorem 3.2.5.

Let us prove the second statement. Fix $\epsilon > 0$ and define the following set for $\phi \in Z(M)$: $\mathrm{CR}_\epsilon(\phi) = \{x :$ there is an ϵ- trajectory from x to x $\}$. Evidently,

$$\mathrm{CR}(\phi) = \bigcap_{\epsilon>0} \mathrm{CR}_\epsilon(\phi).$$

Now denote by $D(\phi)$ the union of the basins of all chain transitive quasi-attractors of ϕ, and for $\epsilon > 0$ denote by $D_\epsilon(\phi)$ the union of the basins of attractors I such that $I \subset \mathrm{CR}_\epsilon(\phi)$. It follows from Lemma 3.1.1 that for any $\epsilon > 0$ the set $D_\epsilon(\phi)$ is open.

Lemma 3.3.1.

$$D(\phi) = \bigcap_{\epsilon>0} D_\epsilon(\phi).$$

Proof. It is evident that for any $\epsilon > 0$ we have $D(\phi) \subset D_\epsilon(\phi)$, hence

$$\bigcap_{\epsilon>0} D_\epsilon(\phi) \supset D(\phi).$$

We claim that

$$\bigcap_{\epsilon>0} D_\epsilon(\phi) \subset D(\phi). \tag{3.18}$$

It suffices to show that

$$\bigcap_{m\geq 0} D_{1/m}(\phi) \subset D(\phi).$$

If

$$x \in \bigcap_{m\geq 0} D_{1/m}(\phi)$$

then for any m there is an attractor I_m such that

$$\omega_x \subset I_m \subset \mathrm{CR}_{1/m}(\phi)$$

(we remind that ω_x is the ω-limit set of $O(x,\phi)$). Hence

$$\omega_x \subset I = \bigcap_{m\geq 0} I_m \subset \mathrm{CR}(\phi),$$

so that $x \in D(\phi)$. This proves (3.18) and our lemma.

Lemma 3.3.2. *If a diffeomorphism ϕ satisfies the STC then*

$$\overline{D_\epsilon(\phi)} = M$$

for all $\epsilon > 0$.

Proof. Consider a simple ordering $>$ on the set of basic sets $\Omega_1, \ldots, \Omega_p$ of ϕ such that

$$\Omega_1 < \ldots < \Omega_p,$$

and $\Omega_i < \Omega_j$ implies $W^u(\Omega_i) \cap W^s(\Omega_j) = \emptyset$.

Let Ω_i be a minimal element with respect to this ordering, that is for all $k \neq i$

$$W^u(\Omega_i) \cap W^s(\Omega_k) = \emptyset.$$

As

$$\bigcup_{j=1}^{p} W^s(\Omega_j) = M \tag{3.19}$$

and

$$W^u(\Omega_i) \cap W^s(\Omega_i) = \Omega_i$$

(see Sect. 0.4) we have

$$W^u(\Omega_i) = \Omega_i.$$

We claim that in this case $W^s(\Omega_i)$ contains a neighborhood of Ω_i, then evidently Ω_i is an attractor. To obtain a contradiction suppose that any neighborhood of Ω_i contains a point x belonging to $W^s(\Omega_k)$, $k \neq i$ (as the number of basic sets is finite we may suppose that k is the same for all x). It follows from Lemma 0.4.1 that in this case

$$W^u(\Omega_i) \cap W^s(\Omega_k) \neq \emptyset,$$

this contradicts the minimality of Ω_i .

Let $\Omega_1, \dots, \Omega_l$ be all minimal elements. As any Ω_i contains a dense trajectory of ϕ , $\Omega_1, \dots, \Omega_l$ are chain transitive attractors. It is easy to see that

$$D(\Omega_i) = W^s(\Omega_i), i = 1, \dots, l.$$

For any $k \in \{l+1, \dots, p\}$ there is $i \in \{1, \dots, l\}$ such that

$$W^u(\Omega_k) \cap W^s(\Omega_i) \neq \emptyset$$

and then

$$W^s(\Omega_k) \subset \overline{W^s(\Omega_i)}$$

(this follows from Lemma 0.4.2). Now (3.19) implies that

$$M = \bigcup_{i=1}^{l} \overline{W^s(\Omega_i)}.$$

Lemma 3.3.3. *Assume that A is a compact ϕ- invariant set such that $A \subset CR_\epsilon(\phi)$ for some $\epsilon > 0$. Then there exists $\epsilon' \in (0, \epsilon)$ and $\alpha > 0$ such that*

$$N_\alpha(A) \subset CR_{\epsilon'}(\phi).$$

Proof. We begin by showing that there is $\epsilon' \in (0, \epsilon)$ such that

$$A \subset \mathrm{CR}_{\epsilon'}(\phi). \tag{3.20}$$

Choose $\delta \in (0, \epsilon/2)$ such that $d(x,y) < \delta$ implies $d(\phi(x), \phi(y)) < \epsilon/2$. Let $X = \{\xi_0, \dots, \xi_m\}$ be a finite, δ- dense subset of A . Now find a finite ϵ- trajectory $\{x_k\}$, $k = 0, \dots, p$, satisfying $x_0 = x_p$, and $X \subset \{x_k\}$. We can do it as follows : $\xi_0, \xi_1 \in A$ hence there exists a finite ϵ-trajectory $\{x_0, \dots, x_r\}$ with $x_0 = \xi_0, \dots, x_r = \xi_1$. After that find a finite ϵ-trajectory $\{x_r, \dots, x_s\}$ with $x_r = \xi_1, x_s = \xi_2$, and so on.

Let

$$\epsilon_0 = \max\{\frac{\epsilon}{2}, \max_{0 \le i \le p-1} d(x_{i+1}, \phi(x_i))\}.$$

By the definition of ϵ-trajectory, $\epsilon_0 < \epsilon$. Define

$$\epsilon' = \frac{\epsilon + \epsilon_0}{2} < \epsilon.$$

Now, if $y, z \in A$ we can choose $x_j, x_i \in X$ to satisfy $d(x_j, \phi(y)) < \delta$ and $d(z, \phi(x_i)) < \delta$. Then $\{y, x_j, \dots, x_i, z\}$ is an ϵ'-trajectory where the ϵ-trajectory from x_j to x_i is obtained from the periodic ϵ-trajectory $\{x_k\}$, $k = 0, \dots, p-1$. Indeed,

$$d(x_j, \phi(y)) < \delta < \epsilon', d(\phi(x_i), z) < \frac{\epsilon}{2} < \epsilon'$$

by the choice of δ . Thus, (3.20) is proved.

Finally, by arguing as in the first part of the proof, we can choose $\alpha \in (0, \epsilon')$ such that $d(x, y) < \alpha$ implies $d(\phi(x), \phi(y)) < \epsilon'$. It follows immediately that

$$N_\alpha(A) \subset \mathrm{CR}_{\epsilon'}(\phi).$$

Now fix arbitrary $\epsilon > 0$ and define the map

$$\Delta_\epsilon : Z(M) \to M^*$$

by

$$\Delta_\epsilon(\phi) = \overline{D_\epsilon(\phi)}.$$

Lemma 3.3.4. *The map Δ_ϵ is lower semi-continuous.*

Proof. Suppose that I is an attractor of ϕ such that

$$I \subset \mathrm{CR}_\epsilon(\phi).$$

By Lemma 3.3.3, there are constants $\epsilon' < \epsilon$ and an arbitrarily small $\alpha > 0$ such that

$$N_\alpha(I) \subset \mathrm{CR}_{\epsilon'}(\phi).$$

We suppose that α is small enough, so that

$$N_\alpha(I) \subset D(I).$$

Take a compact set K such that

$$K \subset D(I), N_\alpha(K) \supset D(I).$$

Use Lemma 3.1.2 to find m with

$$\phi^m(K) \subset N_\alpha(I), \phi^{m+1}(K) \subset N_\alpha(I).$$

There exists a neighborhood W of ϕ such that for any $\psi \in W$ we have

$$\psi^m(K) \subset N_\alpha(I), \psi^{m+1}(K) \subset N_\alpha(I).$$

It follows from Lemma 3.1.3 that any system $\psi \in W$ has an attractor $I(\psi)$ having the properties

$$I(\psi) \subset N_\alpha(I), K \subset D(I(\psi)).$$

Evidently we can take W so small that

$$\mathrm{CR}_{\epsilon'}(\phi) \subset \mathrm{CR}_\epsilon(\psi)$$

for $\psi \in W$. It follows that $I(\psi) \subset \mathrm{CR}_\epsilon(\psi)$, hence $D(I(\psi)) \subset D_\epsilon(\psi)$. It was shown in Remark after the proof of Theorem 3.2.2 that ϕ has at most countably many attractors. As M is compact there are finitely many attractors, $I_1, \dots, I_k$ such that

$$D_\epsilon(\phi) \subset N_\alpha(\bigcup_{j=1}^{k} D(I_j)).$$

Applying the preceding argument to each $I_j, j = 1, \dots, k$, we see that there exists a neighborhood W of ϕ such that any system $\psi \in W$ has attractors $I_j(\psi), j = 1, \dots, k$, with the following properties :

$$D(I_j) \subset N_\alpha(D(I_j(\psi))), I_j(\psi) \subset \mathrm{CR}_\epsilon(\psi),$$

hence,

$$D_\epsilon(\phi) \subset N_\alpha(\bigcup_{j=1}^{k} D(I_j)) \subset$$

$$\subset N_{2\alpha}(\bigcup_{j=1}^{k} D(I_j(\psi))) \subset N_{3\alpha}(D_\epsilon(\psi)),$$

and so

$$\overline{D_\epsilon(\phi)} \subset N_{4\alpha}(\overline{D_\epsilon(\psi)}).$$

As α can be taken arbitrarily small, this proves the lemma.

Now let H_m be the set of continuity points of $\Delta_{1/m}, m = 1, \dots,$ and

$$H = \bigcap_m H_m.$$

It follows from Lemma 3.3.4 that H is a residual subset of $Z(M)$. To complete the proof of Theorem 3.3.1 we now prove the following lemma.

Lemma 3.3.5. *If* $\dim M \leq 3$, *and* $\phi \in H$, *then* $D(\phi)$ *is a residual subset of* M.

Proof. As $\dim M \leq 3$ we can find a sequence of diffeomorphisms $\phi_k \in$STC such that $\lim_{k \to \infty} \phi_k = \phi$ (see Theorem 2.4.1). By Lemma 3.3.2

$$\overline{D_{1/m}(\phi_k)} = M$$

for all k, m. As $\phi \in H$ we have

$$\overline{D_{1/m}(\phi)} = M$$

for all m . Thus any $D_{1/m}(\phi)$ is open and dense in M , so

$$D(\phi) = \bigcap_{m>0} D_{1/m}(\phi)$$

is residual in M (we apply Lemma 3.3.1 here).

Remark. In the original paper [Hu1] M.Hurley published the proved result without the restriction $\dim M \leq 3$. But the proof in [Hu1] used the density of STC in $Z(M)$, and till now this fact is established only in the case $\dim M \leq 3$ (see Theorem 0.1.1).

Now we are going to prove the following result describing a property of Lyapunov stable invariant sets for generic systems in $Z(M)$.

Theorem 3.3.2 [Pi7]. *A generic system $\phi \in Z(M)$ has the property : any Lyapunov stable set I of ϕ is a quasi-attractor of ϕ.*

To prove this theorem take a system $\phi \in Z(M)$, an invariant compact set I of ϕ, and introduce the number

$$\theta(I, \phi) = \sup\{\overline{\lim_{k \to \infty}} d(\phi_k^{t_k}(x_k), I\}.$$

Here we take $\sup \overline{\lim}$ for all sequences of systems ϕ_k, of points x_k , and of numbers t_k such that

$$\lim_{k \to \infty} \phi_k = \phi, \lim_{k \to \infty} x_k \in I, \lim_{k \to \infty} t_k = +\infty.$$

Lemma 3.3.6. *A generic system $\phi \in Z(M)$ has the property : for any Lyapunov stable set I of ϕ*

$$\theta(I, \phi) = 0.$$

Proof. Take a system $\phi \in Z(M)$ having the property described in Theorem 1.3.2. To obtain a contradiction suppose that ϕ has a Lyapunov stable set I with $\theta(I, \phi) > 0$. This implies that there exists $a > 0$ and sequences

$$\phi_k \to \phi, t_k \to +\infty, x_k \to x_0 \in I$$

(we can obtain the last relation passing to a subsequence) with

$$d(\phi_k^{t_k}, I) \geq a.$$

Let ξ be a limit point of $\phi_k^{t_k}(x_k)$, then $d(\xi, I) \geq a$. Apply Lemma 0.3.2 to construct sequences of systems ψ_k and of numbers $\tau_k > 0$ such that

$$\lim_{k \to \infty} \rho_0(\phi_k, \psi_k) = 0,$$

and ξ is a limit point of $\psi_k^{\tau_k}(x_0)$. Then

$$\xi \in P(x_0, \phi). \tag{3.21}$$

As I is Lyapunov stable there is $\delta > 0$ such that for any $x \in N_\delta(I)$ we have

$$O^+(x,\phi) \subset N_{a/2}(I).$$

It follows from the definition of $Q(x_0,\phi)$ that

$$Q(x_0,\phi) \subset \overline{N_{a/2}(I)},$$

hence

$$\xi \notin Q(x_0,\phi). \tag{3.22}$$

The contradiction between (1.23),(3.21), and (3.22) proves the lemma.

Lemma 3.3.7. *Let I be an attractor for a system $\phi \in Z(M)$, and let K be a compact subset of $D(I)$. Then given $\epsilon > 0$ there is a neighborhood $H(K,\epsilon)$ of ϕ in $Z(M)$ such that for any system $\psi \in H(K,\epsilon)$, for any sequences $x_k \in K$, $t_k \to +\infty$*

$$\overline{\lim} d(\psi^{t_k}(x_k), I) < \epsilon.$$

Proof. Take arbitrary $\epsilon > 0$. Find $\delta > 0$ such that

$$\overline{N_\delta(I)} \subset D(I); \phi^k(\overline{N_\delta(I)}) \subset N_{\epsilon/2}(I), k \geq 0.$$

To obtain a contradiction suppose that there exist sequences of systems ϕ_k , of points $x_k \in K$, and of numbers t_k such that

$$\lim_{k\to\infty} \phi_k = \phi, \lim_{k\to\infty} t_k = +\infty, \phi_k^{t_k}(x_k) \notin N_\epsilon(I).$$

Let us consider two cases.

Case 1. $\phi_k^m(x_k) \notin N_\delta(I)$ for $0 \leq m \leq t_k$. Let in this case x_0 be a limit point of x_k , we suppose that $\lim_{k\to\infty} x_k = x_0$. Then for any $m \geq 0$ we have

$$\phi^m(x_0) = \lim_{k\to\infty} \phi_k^m(x_k) \subset M \setminus N_\delta(I).$$

This contradicts to inclusions

$$x_0 \in K \subset D(I).$$

Case 2. For arbitrarily large k there exists $\tau_k \in \{0,\ldots,t_k\}$ such that

$$\xi_k = \phi_k^{\tau_k}(x_k) \in N_\delta(I).$$

Then we can find $\eta_k \in O^+(\xi_k,\phi_k)$ and $\theta_k > 0$ such that $\eta_k \in N_\delta(I)$ and

$$\phi_k^m(\eta_k) \in N_\epsilon(I) \setminus N_\delta(I), 0 \leq m \leq \theta_k, \tag{3.23}$$

$$\phi_k^{\theta_k}(\eta_k) \notin N_\epsilon(I). \tag{3.24}$$

Let η be a limit point of η_k (we suppose $\lim_{k\to\infty}\eta_k = \eta$), then $\eta \in N_\delta(I)$. Hence there exists $\mu > 0$ such that

$$\phi^m(\eta) \in N_\epsilon(I), 0 \leq m \leq \mu; d(\phi^\mu(\eta), I) < \frac{\delta}{2}.$$

It follows that for large k

$$\phi_k^m(\eta_k) \in N_\epsilon(I), 0 \leq m \leq \mu; d(\phi_k^\mu(\eta_k), I) < \frac{\delta}{2}. \tag{3.25}$$

We see that (3.25) contradicts to (3.23), (3.24), this contradiction completes the proof.

Lemma 3.3.8. *A generic system $\phi \in Z(M)$ has the following property. A compact set I being invariant with respect to ϕ is a quasi-attractor of ϕ if and only if $\theta(I,\phi) = 0$.*

Proof. Assume that I is a quasi-attractor of ϕ, so that

$$I = \bigcap_k I_k$$

where I_k are attractors of ϕ , $k = 0, 1, \ldots$. It was mentioned in section 3.1 that we consider in this case attractors I_k such that

$$I_0 \supset I_1 \supset \ldots \supset I_k \supset I_{k+1} \supset \ldots.$$

Take arbitrary $\delta > 0$, and find an attractor I_m with $I_m \subset N_\delta(I)$. Let K be a compact set such that

$$I_m \subset \mathrm{Int} K \subset K \subset N_\delta(I), K \subset D(I_m).$$

Find $\Delta > 0$ with $N_\Delta(I) \subset \mathrm{Int} K$ and $N_\Delta(\phi) \subset H(K,\delta)$ (here $H(K,\delta)$ is a neighborhood of ϕ having the properties described in Lemma 3.3.7). Now consider sequences ϕ_k, x_k, t_k such that

$$\lim_{k\to\infty}\phi_k = \phi, \lim_{k\to\infty} x_k \in I, \lim_{k\to\infty} t_k = +\infty.$$

There exists k_0 such that for $k \geq k_0$ we have

$$\phi_k \in N_\Delta(\phi), x_k \in N_\Delta(I),$$

hence

$$\overline{\lim} d(\phi_k^{t_k}(x_k), I) \leq \delta.$$

We see that for arbitrary $\delta > 0$ we have $\theta(I,\phi) \leq \delta$, so that $\theta(I,\phi) = 0$.

Now assume that I is an invariant compact set for ϕ, and $\theta(I,\phi) = 0$. To show that I is a quasi-attractor we apply here a construction used in [Hu1] for another reason. For natural k define sets Y_k in the following way : $y \in Y_k$ if and only if for some $x \in I$ there exists a $1/k$-trajectory from x to y . It follows immediately from the definition that for any k the set Y_k is open and positively invariant, that is

$$\phi^m(Y_k) \subset Y_k \text{ for } m \geq 0.$$

We claim that

$$N_{1/k}(\phi(Y_k)) \subset Y_k. \tag{3.26}$$

Indeed, if $\xi \in \phi(Y_k)$ then there are $x \in I$, $y \in Y_k$ such that $\{x, \dots, y\}$ is a $1/k$-trajectory. Hence, for any $\eta \in N_{1/k}(\xi)$ the set $\{x, \dots, y, \eta\}$ is a $1/k$-trajectory, so that $\eta \in Y_k$.

It follows from (3.26) that

$$\phi(\overline{Y_k}) \subset Y_k, \tag{3.27}$$

hence Lemma 3.1.3 implies that the set

$$I_k = \bigcap_{m \geq 0} \phi^m(\overline{Y_k})$$

is an attractor of ϕ.

Now let us show that if our system ϕ has the property described in Theorem 1.3.2 then

$$I = \bigcap_{k \geq 1} I_k, \tag{3.28}$$

so that I is a quasi-attractor of ϕ.

Evidently if $x \in I$ then $x \in Y_k$ for any k, and $O^+(x, \phi) \subset Y_k$, hence $x \in I_k$, and

$$x \in \bigcap_{k \geq 1} I_k.$$

To prove that

$$\bigcap_{k \geq 1} I_k \subset I$$

suppose that there exists $y \in (\bigcap_{k \geq 1} I_k) \setminus I$, and let $a = d(y, I) > 0$. As $I_k \subset Y_k$ (see (3.27)), for any k we can find a point $\xi_k \in I$ such that there is a $1/k$-trajectory from ξ_k to y . Consider a point $\xi \in I$ being a limit point of the sequence ξ_k , we suppose that $\lim_{k \to \infty} \xi_k = \xi$. Let

$$\eta_k = \frac{1}{k} + d(\phi(\xi), \phi(\xi_k)),$$

then evidently $\lim_{k \to \infty} \eta_k = 0$, and $\{\xi, \phi(\xi_k), \dots, y\}$ is an η_k-trajectory from ξ to y.

So we obtain that $y \in R(\xi, \phi)$. As for our system ϕ we have $R(\xi, \phi) = P(\xi, \phi)$ it follows that there exist sequences of systems ϕ_k and of numbers t_k such that

$$\lim_{k \to \infty} \phi_k = \phi, y = \phi_k^{t_k}(\xi).$$

It is evident that in this case $\lim_{k\to\infty} t_k = \infty$, hence

$$\theta(I, \phi) \geq a.$$

The contradiction we obtained proves (3.28).

Theorem 3.3.2 follows from Lemmas 3.3.6,3.3.8.

3.4 Stability of Attractors with the STC on the Boundary

It was shown in Theorem 3.2.2 that there exists a residual set C^* in $Z(M)$ such that for any system $\phi \in C^*$ all attractors are stable in $Z(M)$ with respect to R_0 . The method of the proof was not constructive, so that for a given system ϕ it gave no answer to the question : are attractors of ϕ stable in $Z(M)$ with respect to R_0 ? The same is true for the proof of Theorem 3.2.1 in [Hu1].

We are going to show here that some properties usually studied in the theory of structural stability imply stability of attractors in $Z(M)$. Consider a diffeomorphism $\phi : M \to M$ of class C^1 , let I be an attractor of ϕ, and let J be the boundary of I . In Sect. 0.4 we defined the STC on a compact invariant set of a diffeomorphism.

The following result was proved by the author.

Theorem 3.4.1 [Pi3]. *If ϕ is a diffeomorphism satisfying the STC on the boundary J of an attractor I then I is stable in $Z(M)$ with respect to R.*

Later O.Ivanov and the author proved the following theorem being a generalization of Theorem 3.4.1.

Theorem 3.4.2 [I]. *If ϕ is a diffeomorphism satisfying the STC on the boundary J of an attractor I then there exist constants $\Delta, L > 0$ having the property : for any system $\psi \in Z(M)$ with $\rho_0(\phi, \psi) < \Delta$ we can find an attractor $I(\psi)$ such that*

$$R(I, I(\psi)) \leq L\rho_0(\phi, \psi).$$

Before proving Theorem 3.4.2 we state some known results and establish some preliminary lemmas. We assume below that ϕ is a diffeomorphism satisfying the STC on the boundary J of an attractor I. Let $\Omega(J) = \Omega(\phi) \cap J$ (as earlier, $\Omega(\phi)$ is the nonwandering set of ϕ).

In Appendix B of this book we prove that in this case the following statements (B.1)-(B.5) are true :

(B.1) J is an attractor of ϕ;

(B.2) The set $\Omega(J)$ is hyperbolic, and periodic points of ϕ are dense in $\Omega(J)$;

(B.3) For any points $p, q \in \Omega(J)$ the manifolds $W^s(p), W^u(q)$ are transversal;

(B.4) For any point $x \in J$ there exists a point $p \in \Omega(J)$ such that $x \in W^u(p)$;

(B.5) The set $\Omega(J)$ can be decomposed :

$$\Omega(J) = \Omega_1 \cup \ldots \cup \Omega_m, \tag{3.29}$$

here $\Omega_1, \ldots, \Omega_m$ are disjoint compact ϕ-invariant sets, and each of these sets contains a dense trajectory (below in the proof of Theorem 3.4.2 we call $\Omega_1, \ldots, \Omega_m$ basic sets of $\Omega(J)$ or simply basic sets).

We denote in the proof of Theorem 3.4.2 $d(x, J)$ by $d(x)$ for $x \in M$.

Lemma 3.4.1. *There exists a neighborhood V of the set J and numbers $C > 0$,$\lambda \in (0,1)$ such that for any $x \in V$, $k \geq 0$ we have*

$$d(\phi^k(x)) \leq C\lambda^k d(x). \tag{3.30}$$

Proof. The same considerations as in the proof of Lemma 0.4.1 in [Pl2] show that the properties (B.1)-(B.5) of ϕ imply the existence of positive numbers $\alpha_1, \alpha_2, \delta_0$ such that the following statement is true. If $p, q \in \Omega(J)$, $x \in W^u(q)$, and $d(p, x) < \delta_0$ then there exist smooth discs

$$D(\alpha_1, x) \subset W^u(q), D(\alpha_1, p) \subset W^s(p)$$

having a point of transversal intersection, and such that for any $y \in D(\alpha_1, x)$, $z \in D(\alpha_1, p)$ we have

$$\angle(T_y D(\alpha_1, x), T_z D(\alpha_1, p)) \geq \alpha_2$$

(see Sect. 0.4 for definitions).

It is well-known (see [Pl1], Theorem 1.3 of Chap. 4) that if a segment of a trajectory of a diffeomorphism is close to a hyperbolic set then this segment is also hyperbolic (compare with Lemma 2.2.3), and there are analogues of stable and unstable manifolds for it. Let us give a precise statement in the form we need.

There exist numbers $C_0, \alpha_3 > 0$, $\lambda \in (0,1)$ such that the following is true : given $\beta > 0$ we can find $\delta_1 > 0$ such that if $d(\phi^k(x), \Omega(J)) < \delta_1$ for $0 \leq k \leq k_1$, and for a point $p \in \Omega(J)$ we have $d(x, p) < \delta_1$ then there exists a smooth disc $D(\alpha_3, x)$ (we denote it $D^s(\alpha_3, x)$ below) with the properties :

(a) $\dim D^s(\alpha_3, x) = \dim W^s(p)$;

(b) there is a disc $D(\alpha_3, p) \subset W^s(p)$ with $\dim D(\alpha_3, p) = \dim W^s(p)$ such that the C^1-distance between $D^s(\alpha_3, x)$, $D(\alpha_3, p)$ is less than β ;

(c) for $y \in D^s(\alpha_3, x)$, $0 \leq k \leq k_1$ we have

$$d(\phi^k(x), \phi^k(y)) \leq C_0 \lambda^k d(x, y). \tag{3.31}$$

Consider a point $p \in \Omega(J)$, a ball $N_\delta(p)$ where $\delta > 0$ is small enough, and a point $x \in N_\delta(p) \setminus J$. Find a point $z \in J$ such that $d(x) = d(x, z)$ (we remind that $d(x) = d(x, J)$). It follows from (B.4) that there is a point $q \in \Omega(J)$ with $z \in W^u(q)$.

Now it is geometrically evident that if δ is small enough then there exists a point

$$y \in D^s(\alpha_3, x) \cap W^u(q).$$

With δ fixed, we can find $\Gamma_1 > 0$ such that for any three points x, y, z with described properties we have

$$d(x, y) \leq \Gamma_1 d(x, z) \leq \Gamma_1 d(x).$$

As J is an attractor (see (B.1)), any trajectory $O(\xi, \phi)$ such that $\lim_{k \to -\infty} d(\phi^k(\xi)) = 0$, belongs to J. Hence, $W^u(q) \subset J$, and $y \in J$. So for $k \geq 0$

$$d(\phi^k(x)) \leq d(\phi^k(x), \phi^k(y)). \tag{3.32}$$

As $\Omega(J)$ is compact, we can find numbers $\Gamma_0, \delta_2 > 0$ such that for any $x \in N_{\delta_2}(\Omega(J))$ there is a point

$$y \in D^s(\alpha_3, x) \cap J$$

with $d(x, y) \leq \Gamma_0 d(x)$. It follows from (3.31),(3.32) that if

$$d(\phi^k(x), \Omega(J)) < \delta_2 \text{ for } 0 \leq k \leq k_1$$

then

$$d(\phi^k(x)) \leq \Gamma_0 C_0 \lambda^k d(x), 0 \leq k \leq k_1. \tag{3.33}$$

The same ideas as in Remark 1 after Lemma 2.2.7 show that we can find $\delta \in (0, \delta_2)$ such that the neighborhoods $N_\delta(\Omega_j)$, $j = 1, \dots, m$, are unrevisited, that is if for $i_1, i_2 \in \{1, \dots, m\}$, $i_1 \neq i_2$, and for some $x \in M, \kappa > 0$ we have

$$x \in N_\delta(\Omega_{i_1}), \phi^\kappa(x) \in N_\delta(\Omega_{i_2})$$

then

$$\phi^k(x) \notin N_\delta(\Omega_{i_1}) \text{ for } k \geq \kappa.$$

Fix δ having such property. As J is an attractor, we can find a neighborhood V of J such that

(a) $\overline{V} \cap \Omega(\phi) = \Omega(J)$;

(b) $\phi^k(\overline{V}) \subset N_\delta(J)$ for $k \geq 0$.

We claim that this neighborhood V has the desired property. It follows from the classical theorem of G.Birkhoff that there exists $\mu > 0$ such that for any $x \in V$

$$\text{card}\{k \geq 0 : \phi^k(x) \notin N_\delta(\Omega(J))\} \leq \mu.$$

Find $\nu \geq 1$ such that for any x, y

$$d(\phi(x), \phi(y)) \leq \nu d(x, y). \tag{3.34}$$

Let

$$C = (\frac{\nu}{\lambda} C_0 \Gamma_0)^m$$

(here m is the number of different basic sets).

Take $x \in V$. Let us show that for $k \geq 0$ (3.30) is true. It follows from the choice of δ that there exist different integers $i_1, \dots, i_s \in \{1, \dots, m\}$ and integers

$$0 \leq \beta(i_1) \leq \gamma(i_1) < \beta(i_2) \leq \ldots < \beta(i_s) < +\infty$$

such that

$$\phi^k(x) \in N_\delta(\Omega_{i_j}) \text{ for } k \in [\beta(i_j), \gamma(i_j)], j = 1, \dots, s-1;$$

$$\phi^k(x) \in N_\delta(\Omega_{i_s}) \text{ for } k \in [\beta(i_s), +\infty)$$

and

$$\phi^k(x) \notin N_\delta(\Omega(J)) \text{ for } \gamma(i_j) < k < \beta(i_{j+1}),$$

$j = 1, \dots, s-1$. It follows from (3.33), (3.34) that

$$d(\phi^k(x)) \leq \nu^k d(x) \text{ for } 0 \leq k \leq \beta(i_1);$$

$$d(\phi^k(x)) \leq C_0 \Gamma_0 \lambda^{k-\beta(i_1)} d(\phi^{\beta(i_1)}(x))$$

for $\beta(i_1) \leq k \leq \gamma(i_1)$;

$$d(\phi^k(x)) \leq \nu^{k-\gamma(i_1)} d(\phi^{\gamma(i_1)}(x))$$

for $\gamma(i_1) \leq k \leq \beta(i_2)$, and so on. Taking into account that $s \leq m$ and that

$$\beta(i_1) + \beta(i_2) - \gamma(i_1) + \cdots + \beta(i_s) - \gamma(i_{s-1}) \leq \mu$$

we see that (3.30) is true. This completes the proof.

We introduced in the proof a number $\nu \geq 1$ such that (3.34) is true. We suppose that for this ν we also have

$$d(\phi^{-1}(x), \phi^{-1}(y)) \leq \nu d(x, y). \tag{3.35}$$

Find a natural m such that

$$\nu \lambda^{m-1} < 1 \tag{3.36}$$

(here λ is taken from Lemma 3.4.1), and define on V the function

$$v(x) = \sum_{k=0}^{\infty} d^m(\phi^k(x)).$$

Let

$$A = \frac{C^m}{1-\lambda^m}, q = 1 - \frac{1}{A},$$

evidently, $A > 1$ and $q \in (0,1)$.

Lemma 3.4.2. *For $x, y \in V$ we have*

$$(a)\ d^m(x) \le v(x) \le Ad^m(x); \tag{3.37}$$

$$(b)\ |v(x) - v(y)| \le L_1 \max\{d^{m-1}(x), d^{m-1}(y)\} d(x,y), \tag{3.38}$$

where

$$L_1 = mC^{m-1} \sum_{k=0}^{\infty} (\nu\lambda^{m-1})^k;$$

$$(c)\ v(\phi(x)) \le qv(x). \tag{3.39}$$

Proof. Inequalities (3.37) are an immediate corollary of Lemma 3.4.1. To prove (3.38) let us estimate

$$|v(x) - v(y)| \le \sum_{k=0}^{\infty} |d^m(\phi^k(x)) - d^m(\phi^k(y))| \le$$

$$\le \sum_{k=0}^{\infty} d(\phi^k(x), \phi^k(y)) m \max\{d^{m-1}(\phi^k(x)), d^{m-1}(\phi^k(y))\} \le$$

$$\le mC^{m-1} \max\{d^{m-1}(x), d^{m-1}(y)\} \sum_{k=0}^{\infty} (\nu\lambda^{m-1})^k d(x,y)$$

(we note that $d(\phi^k(x), \phi^k(y)) \le \nu^k d(x,y)$ and take (3.30) into account here).

Inequalities (3.37) and evident formula $v(\phi(x)) = v(x) - d^m(x)$ imply (3.39).

Inequality (3.39) means that we can consider $v(x)$ as a Lyapunov function for ϕ in a neighborhood V of J. Let us investigate behaviour of v along trajectories of a system $\psi \in Z(M)$ which is C^0-close to ϕ.

Fix a number γ_0 such that

$$\{x : d(x) \le \gamma_0^{1/m}\} \subset V,$$

and define the set

$$W = \{x \in V : v(x) < \gamma_0\}.$$

It follows from (3.39) that

$$\phi(\overline{W}) \subset W.$$

If we define $W_0 = W \cup I$ then evidently

$$\phi(\overline{W}_0) \subset W_0 \text{ , and } I = \bigcap_{k \ge 0} \phi^k(\overline{W}_0).$$

Find $\Delta_0 > 0$ such that if for $\psi \in Z(M)$ we have

$$\rho_0(\phi, \psi) < \Delta_0 \tag{3.40}$$

then

$$\psi(\overline{W}) \subset W, \psi(\overline{W}_0) \subset W_0.$$

Apply Remark after Lemma 3.1.3 to show that the set

$$I(\psi) = \bigcap_{k \geq 0} \psi^k(\overline{W}_0)$$

is an attractor for ψ.

Lemma 3.4.3. *There exist constants $L_2 > 0$, $\tilde{q} \in (0,1)$ such that for any system ψ which satisfies (3.40) and for any $x \in W$ with*

$$L_2 \rho_0(\phi, \psi) \leq d(x) \tag{3.41}$$

we have

$$v(\psi(x)) < \tilde{q} v(x). \tag{3.42}$$

Proof. Fix a system ψ satisfying (3.40), and let $\rho = \rho_0(\phi, \psi)$,

$$L_2 = \frac{\nu 2^m L_1}{1 - q}, \tilde{q} = \frac{1 + q}{2}.$$

Take x such that (3.41) holds. It follows from (3.35) that $d(x) \leq \nu d(\phi(x))$, hence

$$d(\phi(x)) \geq \frac{d(x)}{\nu} \geq \frac{L_2 \rho}{\nu} \text{ , and } \rho \leq \frac{\nu}{L_2} d(\phi(x)).$$

Thus, we obtain

$$d(\psi(x)) \leq d(\phi(x)) + \rho \leq$$
$$\leq (1 + \frac{\nu}{L_2}) d(\phi(x)) \leq 2d(\phi(x)).$$

Apply (3.38) to estimate

$$| v(\psi(x)) - v(\phi(x)) | \leq 2^{m-1} L_1 \rho d^{m-1}(\phi(x)).$$

It follows from the definition of v that

$$d^m(\phi(x)) \leq v(x). \tag{3.43}$$

Taking (3.43) and $d(x) \leq \nu d(\phi(x))$ into account we see that

$$| v(\psi(x)) - v(\phi(x)) | \leq 2^{m-1} L_1 \nu \rho \frac{v(x)}{d(x)}.$$

Hence, if (3.41) holds, then

$$v(\psi(x)) = v(\phi(x)) + v(\psi(x)) - v(\phi(x)) \leq$$

$$\leq qv(x) + 2^{m-1}L_1\nu\rho\frac{v(x)}{d(x)} \leq \frac{1+q}{2}v(x).$$

This completes the proof.

Now fix $\gamma > 0$, $\Delta_1 \in (0, \Delta_0)$ such that

$$\overline{N_\gamma(I)} \subset W_0,$$

and for any system ψ with

$$\rho_0(\phi, \psi) < \Delta_1 \tag{3.44}$$

we have

$$(\overline{W}_0 \setminus \psi(\overline{W}_0)) \cap \overline{N_\gamma(I)} = \emptyset. \tag{3.45}$$

It follows from the definition of $I(\psi)$ and from (3.44) that if for a point x

$$d(\psi^k(x), I) < \gamma, k \leq 0, \tag{3.46}$$

then $x \in I(\psi)$. Indeed, (3.46) implies $x \in \overline{W}_0$. If we suppose that $x \notin I(\psi)$ then there is $s > 0$ with

$$x \notin \psi^s(\overline{W}_0). \tag{3.47}$$

Find in this case the minimal s such that (3.47) holds, then

$$\psi^{-s+1}(x) \in \overline{W}_0 \setminus \psi(\overline{W}_0)$$

which contradicts to (3.45),(3.46).

Now choose

$$L_3 > A^{1/m}\tilde{q}^{-1/m}L_2.$$

Let us suppose that the number Δ_1 satisfies the inequality

$$\nu\Gamma(L_3+1)\Delta_1 < \gamma_0^{1/m}. \tag{3.48}$$

Lemma 3.4.4. *If for a system ψ (3.44) holds then*

$$r(I(\psi), I) \leq L_3\rho_0(\phi, \psi), \tag{3.49}$$

$$r(\overline{IntI}, I(\psi)) \leq L_3\rho_0(\phi, \psi). \tag{3.50}$$

Proof. Define for integer $s \geq 0$ the sets

$$V_s = \{x \in \overline{W} \setminus I : v(x) \geq \tilde{q}^s\gamma_0\}.$$

Let again $\rho = \rho_0(\phi, \psi)$. Suppose that for some s

$$(\frac{\tilde{q}^s\gamma_0}{A})^{1/m} \geq L_2\rho. \tag{3.51}$$

It follows from (3.37) that for $x \in V_s$ we have $Ad^m(x) \geq \tilde{q}^s \gamma_0$, hence $d(x) \geq L_2 \rho$ and

$$N_{L_2\rho}(x) \cap I = \emptyset.$$

Set $x_k = \psi^{-k}(x)$, $k = 0, 1, \ldots$. As

$$d(x_1, I) = d(\psi^{-1}(x), I) \geq d(\phi^{-1}(x), I) - \rho \geq$$

$$(\frac{L_2}{\nu} - 1)\rho > 0,$$

we see that $x_1 \notin I$. Now we apply (3.42) to prove that if $x \in V_s$ and (3.51) holds then either $x_1 \in V_{s-1}$ or $x_1 \notin \overline{W}_0$.

Let us prove (3.49). Consider a system ψ which satisfies (3.44). Find an integer $p \geq 0$ such that

$$\log L_3 + \log \rho \geq \frac{1}{m}(p \log \tilde{q} + \log \gamma_0) \geq \log(A^{1/m} L_2) + \log \rho \tag{3.52}$$

(we can find such p due to the choice of L_3). Consider a point $x \notin I$ such that $d(x) \geq L_3 \rho$. We claim that $x \notin I(\psi)$. It follows from (3.52) that for $s = 0, \ldots, p$ (3.51) holds. Hence either one of the points x_k does not belong to $\overline{W}_0$ or we have

$$x_0 \in V_p, x_1 \in V_{p-1}, \ldots, x_{p-1} \in V_1 \subset \overline{W}_0 \setminus \psi(\overline{W}_0).$$

In both cases $x \notin I(\psi)$. So, (3.49) is proved.

To prove (3.50) define for $s \geq 0$ the sets

$$V_s^- = \{x \in \text{Int} I \cap \overline{W} : v(x) \geq \tilde{q}^s \gamma_0\}$$

and use similar considerations. Note in this case that if

$$x \in \text{Int} I \setminus \overline{W}, d(x) \geq L_3 \rho$$

then either

$$\psi^{-1}(x) \in \text{Int} I \setminus \overline{W},$$

or $\psi^{-1}(x) \in V_p^-$ with p satisfying (3.52) (we take (3.48) into account here).

It follows from results of V.Pliss obtained in [Pl3] that for our diffeomorphism ϕ there exist numbers $\Delta_2, L_4 > 0$ having the following property : if $\psi \in Z(M)$ is a dynamical system with $\rho_0(\phi, \psi) < \Delta_2$ then for any $y \in J$ there is $x \in M$ such that

$$d(\phi^k(y), \psi^k(x)) \leq L_4 \rho_0(\phi, \psi), k \in \mathbf{Z}. \tag{3.53}$$

To prove this statement one can apply arguments close to ones used in the proof of Theorem A.1 (see Appendix A of this book).

Find $\Delta_3 \in (0, \Delta_2)$ such that $L_4 \Delta_3 < \gamma$ (note that γ is defined after Lemma 3.4.3). Now take a point $y \in J$ and a system $\psi \in Z(M)$ with $\rho_0(\phi, \psi) < \Delta_3$. Let

for a point $x \in M$ (3.53) holds, then $O^-(x,\psi)$ satisfies (3.46). Hence, $x \in I(\psi)$. So, we obtain that

$$r(J, I(\psi)) \le L_4\rho_0(\phi,\psi). \tag{3.54}$$

Take $\Delta = \min(\Delta_1, \Delta_3)$, $L = \max(L_3, L_4)$. Inequalities (3.49), (3.50), (3.54) imply that for $\psi \in Z(M)$ with $\rho_0 \le \Delta$ we have

$$R(I, I(\psi)) \le L\rho_0(\phi,\psi).$$

This completes the proof of Theorem 3.4.2.

V.Pogonysheva generalized Theorem 3.4.1 in the following way.

Theorem 3.4.3 [Po1]. *If ϕ is a diffeomorphism satisfying the STC on the boundary J of an attractor I then I is stable in $Z(M)$ with respect to R_0.*

Proof. Fix $\epsilon > 0$. It follows from Lemma 3.2.1 that there exists $\delta > 0$ such that $\delta < \epsilon$ and

$$r(\overline{M \setminus I}, M \setminus N_\delta(I)) < \epsilon. \tag{3.55}$$

Apply Theorem 3.4.2 to find a neighborhood Y_1 of ϕ in $Z(M)$ such that for any system $\psi \in Y_1$ we have

$$R(I(\psi), I) < \delta. \tag{3.56}$$

Here $I(\psi)$ is the attractor of ψ defined by

$$I(\psi) = \bigcap_{k \ge 0} \psi^k(\overline{W}_0)$$

in the proof of Theorem 3.4.2 before Lemma 3.4.3. It follows from (3.56) that

$$I(\psi) \subset N_\delta(I),$$

hence

$$M \setminus N_\delta(I) \subset M \setminus I(\psi),$$

so, taking (3.55) into account we see that

$$r(\overline{M \setminus I}, \overline{M \setminus I(\psi)}) < \epsilon. \tag{3.57}$$

Let us now show that there exists a neighborhood Y_2 of ϕ such that

$$r(\overline{M \setminus I(\psi)}, \overline{M \setminus I}) < \epsilon \tag{3.58}$$

for $\psi \in Y_2$. To do this we need the following

Lemma 3.4.5. *There exists a neighborhood Y_2 of ϕ such that for $\psi \in Y_2$ we have*

$$(IntI) \setminus N_{\epsilon/2}(J) \subset I(\psi). \tag{3.59}$$

Proof. Let $\rho = \rho_0(\phi,\psi) < \min(\Delta_0, \Delta_1)$ (see (3.40),(3.44)). Take

$$L_5 = A(\nu L_2 + 1)^m$$

(see the proof of Theorem 3.4.2). Fix $r > 0$ such that

$$L_5^{1/m} r < \frac{\epsilon}{2},$$

and let

$$Y_2 = \{\psi \in Z(M) : \rho_0(\phi, \psi) < r\}.$$

To obtain a contradiction suppose that for some system $\psi \in Y_2$ (3.59) is not true, and take a point

$$\xi \in (\text{Int} I) \setminus N_{\epsilon/2}(J), \xi \notin I(\psi). \tag{3.60}$$

Let $\xi_k = \psi^{-k}(\xi)$, $k = 0, 1, \dots$. We claim that if $\xi_k \in I$ for all $k \geq 0$ then $\xi \in I(\psi)$. Indeed, as $I \subset \overline{W}_0$, inclusions $\xi_k \in I$ imply

$$\xi = \psi^k(\xi_k) \in \psi^k(\overline{W}_0),$$

so that

$$\xi \in \bigcap_{k \geq 0} \psi^k(\overline{W}_0) = I(\psi).$$

Hence there is $\kappa \geq 0$ such that $\xi_\kappa \in I$, $\eta = \xi_{\kappa+1} \notin I$. Then

$$d(\eta, I) \leq d(\psi^{-1}(\xi_\kappa), \phi^{-1}(\xi_\kappa)) < \rho$$

(we take into account here that $\phi^{-1}(I) = I$). We can take r so small that $N_r(J) \subset V$. Evidently

$$d(\eta, J) = d(\eta, I) < \rho < r,$$

so we obtain from (3.37) that

$$v(\eta) \leq A\rho^m \leq L_5 \rho^m.$$

Let us show that $\psi \in Y_2$ maps the set

$$V^* = \{x \in V : v(x) \leq L_5 \rho^m\}$$

into itself. Indeed, if $d(x, J) \geq L_2 \rho$ then it follows from Lemma 3.4.3 that $v(\psi(x)) < v(x)$. If $d(x, J) < L_2 \rho$ then

$$d(\psi(x), J) \leq d(\phi(x), J) + \rho \leq \nu d(x, J) + \rho \leq$$

$$\leq (\nu L_2 + 1)\rho$$

Hence,

$$[v(\psi(x))]^{1/m} \leq A^{1/m} d(\psi(x), J) \leq A^{1/m}(\nu L_2 + 1)\rho \leq L_5^{1/m}\rho,$$

and again

$$v(\psi(x)) \leq L_5 \rho^m.$$

As $\eta \in V^*$ we obtain that for $k \geq 0$ we have $\psi^k(\eta) \in V^*$. But as $\xi = \psi^\kappa(\eta)$ it follows that

$$v(\xi) \leq L_5 \rho^m,$$

and

$$d(\xi, J) \leq [v(\xi)]^{1/m} < L_5^{1/m} r < \frac{\epsilon}{2}.$$

We obtain a contradiction with (3.60). This proves the lemma.

Now we show that (3.59) implies (3.58). Indeed it follows from (3.59) that

$$M \setminus I(\psi) \subset M \setminus ((\mathrm{Int} I) \setminus N_{\epsilon/2}(J)).$$

Taking into account that

$$(\mathrm{Int} I) \setminus N_{\epsilon/2}(J) = (\mathrm{Int} I) \setminus N_{\epsilon/2}(\partial \mathrm{Int} I)$$

we have

$$M \setminus I(\psi) \subset (M \setminus \mathrm{Int} I) \cup N_{\epsilon/2}(\partial \mathrm{Int} I).$$

It is geometrically evident that

$$\overline{M \setminus I(\psi)} \subset (M \setminus \mathrm{Int} I) \cup \overline{N_{\epsilon/2}(\partial \mathrm{Int} I)}. \tag{3.61}$$

But the set in the right hand side of (3.61) is a subset of $N_\epsilon(\overline{M \setminus I})$. Indeed, if a point x belongs to $M \setminus \mathrm{Int} I$ then either $x \in M \setminus I$ or $x \in J$. In both cases $x \in \overline{M \setminus I}$. As $\partial \mathrm{Int} I \subset \partial(M \setminus I)$ we see that

$$\overline{N_{\epsilon/2}(\partial \mathrm{Int} I)} \subset N_\epsilon(\overline{M \setminus I}),$$

so (3.58) is established. Taking $Y = Y_1 \cap Y_2$ we see that for any system $\psi \in Y$

$$R_0(I, I(\psi)) < \epsilon.$$

This completes the proof of Theorem 3.4.3.

3.5 Stability of Attractors for Morse-Smale Diffeomorphisms

It follows from theorems of the previous section that if a diffeomorphism ϕ satisfies the STC then any attractor of ϕ is stable in $Z(M)$ with respect to metrics R, R_0 . Here we describe some results of V.Pogonysheva [Po1] devoted to stability of attractors with respect to R_1, R_2 . V.Pogonysheva studied in [Po1] stability of attractors for Morse-Smale diffeomorphisms - that is for simplest diffeomorphisms satisfying the STC. These results show that in contrast to the case of metrics R, R_0 stability with respect to R_1, R_2 depends not only on structural properties of the system but also on the geometry of an attractor.

Let ϕ be a Morse-Smale diffeomorphism of M . We consider here the case when Per(ϕ)=Fix(ϕ). Denote P =Fix(ϕ) . Consider an attractor I of ϕ , let $J = \partial I$, and let $J_1 = \partial \mathrm{Int} I$.

Lemma 3.5.1. *If $p \in P \cap J_1$ then $W^u(p) \subset J_1$.*

Proof. If U is a neighborhood of J then evidently

$$(\overline{M \setminus I}) \cup U$$

is a neighborhood of $\overline{M \setminus I}$. It follows from Theorem B.1 that J is an attractor, hence $\overline{M \setminus I}$ is also an attractor of ϕ. As

$$J_1 = \partial(M \setminus I)$$

we see that $p \in P \cap \partial(\overline{M \setminus I})$. The same Theorem B.1 shows that $\partial(\overline{M \setminus I})$ is an attractor, so that if $x \notin J_1$ then

$$\lim_{k \to -\infty} d(\phi^k(x), J_1) = 0$$

is impossible. This shows that any $x \in W^u(p)$ belongs to J_1 .

Theorem 3.5.1. *If I is an attractor of ϕ such that*

$$I \neq \overline{\mathit{IntI}} \tag{3.62}$$

then I is not stable in $Z(M)$ with respect to R_1 .

Proof. Suppose that (3.62) holds, and consider a point

$$x \in I \setminus \overline{\mathrm{Int} I}. \tag{3.63}$$

As

$$M = \bigcup_{p \in P} W^u(p)$$

there is a fixed point p of ϕ such that $x \in W^u(p)$. It follows from (3.63) that $x \in J = \partial I$. The boundary J of I is invariant, hence $O(x, \phi) \subset J$. As

$$\lim_{k \to -\infty} \phi^k(x) = p$$

and J is closed, we obtain that $p \in J$. Now Lemma 5.3.1 implies that $p \in J \setminus J_1$ (otherwise

$$x \in \partial \mathrm{Int} I \subset \overline{\mathrm{Int} I}).$$

Hence there is a neighborhood V of p such that

$$V \cap \overline{\mathrm{Int} I} = \emptyset. \tag{3.64}$$

Let us introduce the following set T - the set of fixed points $p \in J$ having the property : there exists a neighborhood U of p such that

$$U \cap J = U \cap W^u(p). \tag{3.65}$$

Lemma 3.5.2. $T \neq \emptyset$.

Proof. It was shown that there exists a point $p \in P \cap (J \setminus J_1)$. If $p \notin T$ then there is a sequence of points $x_m \in J \setminus W^u(p)$ such that $\lim_{m \to \infty} x_m = p$. Find for any x_m the point $q_m \in P \cap J$ with

$$x_m \in W^u(q_m).$$

As P is finite there is a point $p_1 \in P \cap J$ and a subsequence x_{m_k} of x_m such that $p \neq p_1$ and

$$x_{m_k} \in W^u(p_1).$$

Then

$$p \in \omega(W^u(p_1)).$$

It follows from Lemma 0.4.4 that in this case

$$W^u(p) \subset \overline{W^u(p_1)}.$$

If $p_1 \in \overline{\mathrm{Int} I}$ then $W^u(p) \subset \overline{\mathrm{Int} I}$ (we apply Lemma 3.5.1 here) and $W^u(p) \subset \overline{\mathrm{Int} I}$, this contradicts to properties of p . Hence, $p_1 \in J \setminus J_1$. Now either $p_1 \in T$ or there exists $p_2 \in J \setminus J_1$ such that

$$W^u(p_1) \subset \overline{W^u(p_2)},$$

and so on. Thus, we obtain

$$p \leftarrow p_1 \leftarrow p_2 \leftarrow \cdots$$

(see Lemma 0.4.4). As the set P is finite and $p, p_1, p_2, \ldots$ are distinct, there is a point $p_m \in T$.

Now we take a point $p \in T$. Find neighborhoods N_1 of I and N_2 of p such that :

(a) $N_2 \subset N_1$, $\overline{N_2} \subset U$ (here U is a neighborhood which has the property (3.65));

(b) $\overline{N_2} \cap \overline{\mathrm{Int} I} = \emptyset$;

(c) $\overline{N_1} \subset D(I)$;

(d) the Grobman-Hartman Theorem [Pi8] is true for N_2.

It follows from (a),(c) that for any $x \in N_2 \setminus W^u(p)$ (hence, for $x \in N_2 \setminus J$) we have $x \notin I$, so that there is $k < 0$ with $\phi^k(x) \notin N_1$.

Take $\epsilon_0 > 0$ such that

$$\epsilon_0 \leq \min(\operatorname{diam} N_2, \min_{x \in N_2} d(x, \overline{\operatorname{Int} I})),$$

$$I \subset N_{\epsilon_0}(I) \subset N_1 \subset \overline{N_1} \subset D(I).$$

The same ideas as in the proof of Theorem 3.4.2 show that there is a neighborhood K_0 of ϕ such that for any system $\psi \in K_0$ the set

$$I(\psi) = \bigcap_{k \geq 0} \psi^k(N_1)$$

is an attractor of ψ ,

$$I(\psi) \subset N_{\epsilon_0}(I) \subset \overline{N_1} \subset D(I(\psi)),$$

and

$$R_0(I(\psi), I) < \epsilon_0.$$

It is easy to see that $I(\psi)$ is the maximal attractor of ψ in $N_{\epsilon_0}(I)$, that is for any attractor I^* of ψ with $I^* \subset N_{\epsilon_0}(I)$ we have $I^* \subset I(\psi)$. Indeed, if there is $x \in I^* \setminus I(\psi)$ then we can find $m \in \mathbf{Z}$ such that $\psi^m(x) \notin \overline{N_1}$ (we apply here corollary of Lemma 3.1.2). This contradicts to the invariance of I^* .

We claim that for any neighborhood K of ϕ in $Z(M)$ there is a system $\psi \in K$ such that for any attractor I^* of ψ either

$$r(I, I^*) \geq \epsilon_0, \tag{3.66}$$

or

$$R(\overline{\operatorname{Int} I}, \overline{\operatorname{Int} I^*}) \geq \epsilon_0, \tag{3.67}$$

hence I is not stable in $Z(M)$ with respect to R_1.

Take $K \subset K_0$. As the Grobman-Hartman Theorem is true for N_2 there exists a homeomorphism h mapping N_2 onto a neighborhood D of the origin in $\mathbf{R}^n$ and such that the following holds. If $x \in N_2$ and $\phi(x) \in N_2$ then

$$h(\phi(x)) = \Phi(h(x))$$

where $\Phi(y) = Ly$, here y is the coordinate in $\mathbf{R}^n$, and L is a non-singular $n \times n$ matrix (indeed L is the matrix of $D\phi(p)$ in local coordinates). We can choose coordinates in $\mathbf{R}^n$ so that L has its Jordan canonical form.

We consider below the case when the matrix L is diagonal. This can be done by arbitrarily C^1-small perturbation of ϕ. As our diffeomorphism ϕ is structurally stable, for any diffeomorphism $\tilde{\phi}$, C^1-close to ϕ, there is a homeomorphism H close to the identity which topologically conjugates ϕ and $\tilde{\phi}$. It is evident that the attractor I of ϕ is stable in $Z(M)$ with respect to any metric S on M^* introduced in Sect. 0.2 if and only if the attractor $H(I)$ of $\tilde{\phi}$ is stable with respect to S.

So, in our case

$$L = \operatorname{diag}\{\lambda_1, \ldots, \lambda_n\}.$$

Take a function $\alpha : \mathbf{R} \to \mathbf{R}$, $\alpha \in C^\infty(\mathbf{R})$, such that

$$\alpha(t) = \begin{cases} 0 & \text{for } t \leq 1; \\ 1 & \text{for } t \geq 2; \end{cases}$$

$$0 < \alpha(t) < 1 \text{ for } t \in (1,2).$$

Fix $\Delta > 0$, and consider the matrix $L_\Delta(y)$ which is diagonal, and its diagonal element with index i equals to

$$\alpha(\frac{y^2}{\Delta})\lambda_i + (1 - \alpha(\frac{y^2}{\Delta}))\text{sign}\lambda_i$$

(here $y^2 = y_1^2 + \cdots + y_n^2$).

Let

$$D_0 = \{y \in \mathbf{R}^n : y^2 \leq \Delta\}, D_1 = \{y \in \mathbf{R}^n : y^2 \leq 2\Delta\}.$$

Then evidently $L_\Delta(y) = L$ for $y \in \mathbf{R}^n \setminus D_1$, and for $y \in D_0$ the matrix $L_\Delta(y)$ is a diagonal matrix for which any diagonal element equals either to 1 or to -1. If for a matrix $A = \{a_{ij}\}$ we let

$$\| A \|_1 = \max_{i,j} | a_{ij} |$$

then for $y \in \mathbf{R}^n$ we have

$$\| L - L_\Delta(y) \|_1 \leq \| L \|_1 + 1. \tag{3.68}$$

Define $\Psi : \mathbf{R}^n \to \mathbf{R}^n$ by $\Psi(y) = L_\Delta(y)y$. It follows from (3.68) that given a neighborhhod K of ϕ in $Z(M)$ we can find $\Delta > 0$ such that :

(a) the system $\psi = h^{-1} \circ \Psi \circ h$ is in K ;

(b) $\phi(x) = \psi(x)$ for $x \notin U_2$.

As $L_\Delta^2(y) = E_n$ for $y \in D_0$, any point belonging to the set

$$N_3 = h^{-1}(D_0)$$

is either a fixed point of ψ or a periodic point of ψ of period 2.

Let B^+ be the subspace of $\mathbf{R}^n$ spanned by the eigenvectors which correspond to eigenvalues λ_j of L such that $| \lambda_j |> 1$. Then evidently

$$h(U_2 \cap W^u(p)) = h(U_2) \cap B^+.$$

Notice that for $y \in B^+ \setminus D_0$ we have

$$\lim_{k \to -\infty} d(\Psi^k(y), D_0) = 0.$$

Hence if $q \in J \cap N_2$ then

$$\lim_{k \to -\infty} d(\psi^k(q), N_3) = 0.$$

Now consider an attractor I^* of ψ. If $I^* \cap N_2 = \emptyset$ then evidently (3.66) holds. Now suppose that $I^* \cap N_2 \neq \emptyset$. We consider different cases.

Case 1. $q \in N_3$, so that $I^* \cap N_3 \neq \emptyset$.

Lemma 3.5.3. *If $I^* \cap N_3 \neq \emptyset$ then $N_3 \subset I^*$.*

Proof. Take $x \in I^* \cap N_3$. Let is show that there exists a neighborhood Q of x such that

$$Q \cap N_3 \subset I^*. \tag{3.69}$$

To obtain a contradiction suppose that for any neighborhood Q of x

$$(Q \cap N_3) \setminus I^* \neq \emptyset. \tag{3.70}$$

Find in this case a neighborhood Q such that (3.70) holds, and

$$\overline{Q} \subset D(I^*).$$

Take $y \in Q \cap N_3$, $y \notin I^*$. Apply corollary of Lemma 3.1.2 to find m_0 such that $\psi^m(y) \notin \overline{Q}$ for $m \leq m_0$. This contradicts to $\psi^{2m}(y) = y$.

As N_3 is a homeomorphic image of the ball D_0 this proves the lemma.

It follows from Lemma 3.5.3 that in the first case $p \in \mathrm{Int}I^*$, so that (3.67) holds.

Case 2. $q \in J$. It was shown earlier that in this case $\lim_{k\to-\infty} d(\psi^k(q), N_3) = 0$, hence $N_3 \cap I^* \neq \emptyset$, so again $p \in \mathrm{Int}I^*$.

Case 3. For any $m < 0$ we have

$$\psi^m(q) \in N_2.$$

It follows from the construction of ψ that if r is a limit point of $O^-(q,\psi)$ then $r \in N_3$. So in this case again $N_3 \cap I^* \neq \emptyset$, and $p \in \mathrm{Int}I^*$. Now it remains to consider

Case 4. $q \notin J$, and for some $m < 0$ we have

$$\psi^m(q) \notin N_2.$$

By the choice of N_1, N_2 in this case there is $k < 0$ such that

$$\phi^k(\psi^m(q)) = \psi^{m+k}(q) \notin N_1.$$

This contradicts to inclusions

$$q \in I^* \subset I(\psi) \subset \overline{N_1}.$$

So, we showed that I is not stable in $Z(M)$ with respect to R_1 .

Theorem 3.5.2. *If I is an attractor of an arbitrary system $\phi \in Z(M)$ such that*

(a) $\overline{\mathrm{Int}I} = I$;

(b) I is stable in $Z(M)$ with respect to R_0 ,

then I is stable in $Z(M)$ with respect to R_2

Proof. Fix arbitrary $\epsilon > 0$. Apply Lemma 3.2.1 to find $\delta \in (0, \epsilon)$ such that for $K = \overline{M \setminus I}$ we have

$$\overline{M \setminus K} \subset N_\epsilon(M \setminus N_\delta(K)).$$

As

$$\overline{\mathrm{Int} I} = M \setminus \overline{\overline{(M \setminus I)}}$$

we obtain that

$$\overline{\mathrm{Int} I} \subset N_\epsilon(M \setminus N_\delta(\overline{M \setminus I})). \tag{3.71}$$

Our attractor I is stable with respect to R_0. Find a neighborhood K of ϕ in $Z(M)$ such that for any $\psi \in K$ there is an attractor $I(\psi)$ with

$$R_0(I, I(\psi)) < \delta. \tag{3.72}$$

As $\delta < \epsilon$ it follows from (3.72) and from the assumption (a) that

$$r(\overline{\mathrm{Int} I(\psi)}, \overline{\mathrm{Int} I}) \leq r(I(\psi), I) < \epsilon. \tag{3.73}$$

Note that (3.72) implies

$$\overline{M \setminus I(\psi)} \subset N_\delta(\overline{M \setminus I}),$$

hence

$$M \setminus N_\delta(\overline{M \setminus I}) \subset \overline{\mathrm{Int} I(\psi)}.$$

This last inclusion and (3.71) imply

$$\overline{\mathrm{Int} I} \subset N_\epsilon(\overline{\mathrm{Int} I(\psi)}). \tag{3.74}$$

Now we obtain from (3.72) - (3.74) that

$$R_2(I, I(\psi)) < \epsilon.$$

The following result is a corollary of Theorems 3.4.3, 3.5.1, and 3.5.2.

Theorem 3.5.3. *Assume that ϕ is a Morse-Smale diffeomorphism with Per(ϕ) =Fix(ϕ). An attractor I of ϕ is stable in $Z(M)$ with respect to R_2 if and only if*

$$I = \overline{\mathit{Int} I}.$$

Remark. An analogous statement is true in the case of an arbitrary Morse-Smale diffeomorphism.

4. Limit Sets of Domains

4.1 Lyapunov Stability of Limit Sets

Consider a dynamical system $\phi \in Z(M)$ and a subset $G \subset M$. We define the ω - *limit set* of G in the system ϕ in the following way:

$$\omega(G,\phi) = \{y = \lim_{k\to\infty} \phi^{t_k}(x_k) : x_k \in G, \lim_{k\to\infty} t_k = +\infty\}.$$

Below we usually say "limit set" instead of "ω - limit set", and write $\omega(G)$ instead of $\omega(G,\phi)$ (we use this last notation only when we study the dependence of $\omega(G,\phi)$ on ϕ).

It is easy to see that for any ϕ,G the set $\omega(G)$ is ϕ - invariant and compact, and that $\omega(\overline{G}) = \omega(G)$.

The motivation for study of Lyapunov stability of limit sets was based on the following conjecture of V.Arnold posed in a private conversation with the author.

Consider, for example, the flow on $\mathbf{R}^n$ which is generated by a differential equation

$$\frac{dx}{dt} = f(x)$$

where $f(0) = f(1) = 0$, and $f(x) < 0$ for $x \neq 0,1$ (see Fig. 4.1). It is easy to understand that if we take $G = [2,3]$ then $\omega(G)$ (we don't give a definition for this geometrically evident object in the case of a flow) is the rest point $x = 1$. For arbitrary neighborhood U of the rest point $x = 1$ the set $\omega(U)$ coincides with the segment $[0,1]$, so that this set is considerably different from $\omega(G)$. It was conjectured by V.Arnold that the described phenomenon is "impossible on a physical level". One understands immediately that if for a set G and for any small neighborhood U of $\omega(G)$ the sets $\omega(G)$, $\omega(U)$ are considerably different then the set $\omega(G)$ is Lyapunov unstable. So it is important to be able to establish Lyapunov stability of limit sets.

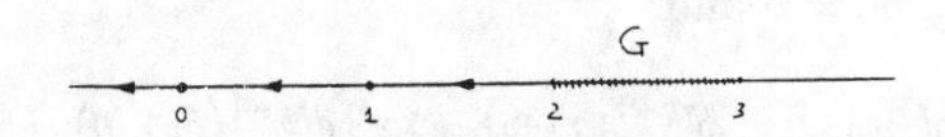

Figure 4.1

We begin with investigation of limit sets for one-parameter monotone families of subsets of M. Consider a family $K_t, t \in (a,b)$, of subsets of M, which satisfies the following condition:

$$\text{-for any } t, \tau \in (a,b) \text{ with } t < \tau \text{ there exists } \delta > 0 \text{ with } N_\delta(K_t) \subset K_\tau. \quad (4.1)$$

Theorem 4.1.1. *A generic system $\phi \in Z(M)$ has the following property. If $K_t, t \in (a,b)$, is a family of subsets of M which satisfies (4.1), then there is a countable set $B \subset (a,b)$ such that for $t \in (a,b) \setminus B$ the set $\omega(K_t)$ is Lyapunov stable.*

To prove this theorem we need some preliminary results.

Lemma 4.1.1. *If for a system $\phi \in Z(M)$ and for a set $G \subset M$ the set $\omega(G)$ is not Lyapunov stable then there exists a point $x \in \overline{G}$ such that*

$$R^\omega(x,\phi) \setminus \omega(G) \neq \emptyset.$$

Proof. As $\omega(G)$ is not Lyapunov stable there exists $\epsilon > 0$ and sequences $\xi_k \in M, t_k \to +\infty$ as $k \to \infty$ such that

$$\lim_{k\to\infty} \xi_k \in \omega(G), d(\phi^{t_k}(\xi_k), \omega(G)) \geq \epsilon.$$

Let ξ, η be limit points of $\xi_k, \phi^{t_k}(\xi_k)$, for simplicity we suppose that

$$\xi_k \to \xi, \phi^{t_k}(\xi_k) \to \eta \text{ as } k \to \infty.$$

Then $d(\eta, \omega(G)) \geq \epsilon > 0$. As $\xi \in \omega(G)$ there is a sequence of points $x_k \in G$ and a sequence $\tau_k \to +\infty$ as $k \to \infty$ such that

$$\lim_{k\to\infty} \phi^{\tau_k}(x_k) = \xi.$$

Let $x \in \overline{G}$ be a limit point of x_k, we suppose that

$$\lim_{k\to\infty} x_k = x.$$

Denote

$$\delta_k = \max\{d(\phi(x), \phi(x_k)), d(\phi^{\tau_k}(x_k), \xi_k), d(\phi^{t_k}(\xi_k), \eta)\}.$$

Then evidently $\lim_{k\to\infty} \delta_k = 0$ and

$$(x, \phi(x_k), \ldots, \phi^{\tau_k - 1}(x_k), \xi_k, \ldots, \phi^{t_k - 1}(\xi_k), \eta)$$

is a finite δ_k - trajectory from x to η. As $\lim_{k\to\infty} \tau_k + t_k = \infty$ we obtain that $\eta \in R^\omega(x,\phi)$. So, $\eta \in R^\omega(x,\phi) \setminus \omega(G)$.

Now consider a family $K_t, t \in (a,b)$, which has the property (4.1) . Define the map $\Phi : (a,b) \to M^*$ by

$$\Phi(t) = \omega(K_t).$$

Lemma 4.1.2. *A generic system $\phi \in Z(M)$ has the following property: if $t \in (a,b)$ is a continuity point of Φ (with respect to Hausdorff metric R on M^*) then the set $\omega(K_t)$ is Lyapunov stable.*

Proof. Let Z^* be a residual subset of $Z(M)$ such that for any $x \in M$ (1.23) and (1.28) hold. Take $\phi \in Z^*$. Now suppose that for some $t_0 \in (a,b)$ the set

$$\omega(K_{t_0}) = \Phi(t_0)$$

is not Lyapunov stable. Apply Lemma 4.1.1 to find a point $x \in \overline{K_{t_0}}$ and a point $\eta \in R^\omega(x,\phi)$ such that

$$\eta \notin \Phi(t_0).$$

As $\phi \in Z^*$ we have

$$R^\omega(x,\phi) = Q^\omega(x,\phi),$$

hence there exists a sequence of points $x_k \in M$ and a sequence $t_k \to +\infty$ such that

$$\lim_{k\to\infty} x_k = x, \lim_{k\to\infty} \phi^{t_k}(x_k) = \eta.$$

Let t_m be a sequence in (a,b) such that

$$t_m > t_0, \lim_{m\to\infty} t_m = t_0.$$

Find for any m a corresponding $\delta(m) > 0$ such that

$$N_{\delta(m)}(K_{t_0}) \subset K_{t_m}.$$

As $x \in \overline{K_{t_0}}$, for fixed m there is $k(m)$ with

$$x_k \in N_{\delta(m)}(K_{t_0}) \subset K_{t_m}$$

for $k \geq k(m)$. It follows immediately that for any m we have

$$\eta \in \omega(K_{t_m}) = \Phi(t_m).$$

Evidently

$$R(\Phi(t_m), \Phi(t_0)) \geq d(\eta, \Phi(t_0)) > 0,$$

hence t_0 is not a continuity point of Φ. This proves the lemma.

Let us say that a map $F : (\alpha,\beta) \to M^*$ is increasing if for any $t, \tau \in (\alpha,\beta)$ with $\tau < t$ we have

$$F(t) \subset F(\tau).$$

N.Scherbina established the following result.

Lemma 4.1.3 [Sc]. *If $F : (\alpha, \beta) \to M^*$ is an increasing map then there exists a countable subset $B \subset (\alpha, \beta)$ such that any $t \in (\alpha, \beta) \setminus B$ is a continuity point of F.*

Proof. Fix $t_0 \in (\alpha, \beta)$, and let

$$H(t_0) = \lim_{\delta \to 0} \sup_{t, \tau \in N_\delta(t_0)} R(F(t), F(\tau)).$$

It is evident that $t_0 \in (\alpha, \beta)$ is a continuity point of F if and only if $H(t_0) = 0$. As the map F is increasing, for any $t_1, t_2, t_3 \in (\alpha, \beta)$ with $t_1 \le t_2 \le t_3$ we have

$$R(F(t_1), F(t_2)) \le R(F(t_1), F(t_3)).$$

Let us show that F has the following property
(P): if $H(t_0) = a > 0$ then for any t_1, t_2 with $t_1 < t_0 < t_2$

$$R(F(t_1), F(t_2)) \ge a.$$

Indeed, for arbitrary $\epsilon > 0$ we can find τ_1, τ_2 such that $t_1 < \tau_1 < \tau_2 < t_2$, and

$$R(F(\tau_1), F(\tau_2)) \ge a - \epsilon.$$

We can estimate

$$R(F(t_1), F(t_2)) \ge R(F(t_1), F(\tau_2)) \ge R(F(\tau_1), F(\tau_2)) \ge a - \epsilon.$$

As ϵ is arbitrary, this proves (P).

Let B denote the set of discontinuity points of F. Let for integer $m \ge 1$

$$B_m = \{t_0 \in (\alpha, \beta) : H(t_0) \ge \frac{1}{m}\}.$$

Evidently

$$B = \bigcup_{m \ge 1} B_m.$$

We shall show that every set B_m is finite, this will prove the lemma. Fix m, let

$$\epsilon_m = \frac{1}{4m},$$

and consider a finite ϵ_m - net $\{x_1, \ldots, x_p\}$ for M. For $t \in (\alpha, \beta)$ denote

$$\Psi(t) = N_{\epsilon_m}(F(t)),$$

and let

$$\tau_k = \inf\{t \in (\alpha, \beta) : x_k \in \Psi(t)\}, k = 1, \ldots, p,$$

(if a point $x_k \notin \Psi(t)$ for all $t \in (\alpha, \beta)$ set $\tau_k = \beta$).

Our map F is increasing, hence Ψ is also increasing, and for $t \in (\tau_k, \beta)$ we have $x_k \in \Psi(t)$.

We claim that

$$B_m \subset \{\tau_1, \ldots, \tau_p\}. \tag{4.2}$$

Fix $t_0 \in B_m$, and $t_1, t_2 \in (\alpha, \beta)$ such that $t_1 < t_0 < t_2$. As $H(t_0) \geq \frac{1}{m}$ we apply the property (P) to show that

$$R(F(t_1), F(t_2)) \geq \frac{1}{m}.$$

Hence there exists a point $z \in F(t_2)$ such that

$$\inf_{x \in F(t_1)} d(z, x) \geq \frac{1}{2m}.$$

Find a point $x_r \in \{x_1, \ldots, x_p\}$ with

$$d(x_r, z) < \frac{1}{4m}.$$

As $z \in F(t_2)$ we obtain that

$$x_r \in \Psi(t_2). \tag{4.3}$$

On the other hand,

$$\inf_{x \in F(t_1)} d(x_r, x) \geq \inf_{x \in F(t_1)} d(z, x) - d(x_r, z) \geq \frac{1}{4m},$$

so that

$$x_r \notin \Psi(t_1). \tag{4.4}$$

It follows from (4.3),(4.4) that

$$t_1 \leq \tau_r \leq t_2. \tag{4.5}$$

As (t_1, t_2) is an arbitrary segment containing t_0, and the set $\{\tau_1, \ldots, \tau_p\}$ is finite, we obtain from (4.5) that $t_0 = \tau_r$.

This proves (4.2), hence the set B_m is finite.

Now we can complete the proof of Theorem 4.1.1. Fix a system ϕ which has the property described in Lemma 4.1.2, and define the corresponding map $\Phi : (a, b) \to M^*$. It is evident that the map Φ is increasing. Let B be the set of discontinuity points of Φ. It follows from Lemma 4.1.3 that B is countable, and it follows from Lemma 4.1.2 that for $t \in (a, b) \setminus B$ the set $\omega(K_t)$ is Lyapunov stable.

One of the most important examples of families K_t having the property (4.1) is the following one. Fix a point $x \in M$, and let for $r \in (0, +\infty)$ $K_r = N_r(x)$. Evidently if $0 < r_1 < r_2$, and $\delta = r_2 - r_1$ then

$$N_\delta(K_{r_1}) \subset K_{r_2}.$$

So, the following result [Pi7] is a corollary of Theorem 4.1.1. A weak variant of this result was originally published in [Pi4].

Theorem 4.1.2. *A generic system $\phi \in Z(M)$ has the following property. Given $x \in M$ there exists a countable set $B(x)$ such that for any $r \in (0, +\infty) \setminus B(x)$ the set $\omega(N_r(x))$ is Lyapunov stable.*

Now we are going to prove a result obtained by V.Pogonysheva in [Po2].

Theorem 4.1.3. *For a generic system $\phi \in Z(M)$ there is a residual subset L of (M^*, R_0) such that for any $G \in L$ the set $\omega(G)$ is Lyapunov stable.*

Previously we need a simple geometric statement. Let for any $G \subset M^*, t \geq 0$

$$\tilde{N}_t(G) = \{x \in M : d(x, G) \leq t\}.$$

Evidently, $\tilde{N}_0(G) = G$, and any $\tilde{N}_t(G) \in M^*$.

Lemma 4.1.4. *For any $G \in M^*$*

$$\lim_{t \to 0} R_0(\tilde{N}_t(G), G) = 0.$$

Proof. Let $K_t = \tilde{N}_t(G)$ for $t \geq 0$. Evidently,

$$\lim_{t \to 0} R(K_t, G) = 0,$$

so we have only to show that

$$\lim_{t \to 0} R(\overline{M \setminus K_t}, \overline{M \setminus G}) = 0. \tag{4.6}$$

Denote $A_t = M \setminus N_t(G)$, $B_t = \overline{M \setminus K_t}$. For any $t > 0$ and for any $x \notin N_{2t}(G)$ we have $d(x, t) \geq 2t > t$, hence $x \notin K_t$, so we obtain that $A_{2t} \subset B_t$. As we always have $\overline{M \setminus K_t} \subset M \setminus N_t(G)$ we see that for $t > 0$

$$A_{2t} \subset B_t \subset A_t,$$

therefore,

$$R(A_t, B_t) \leq R(A_{2t}, A_t).$$

Fix arbitrary $\epsilon > 0$ and apply Lemma 3.2.1 to find $\tau_0 > 0$ such that

$$R(A_\tau, \overline{M \setminus G}) < \frac{\epsilon}{3} \text{ for } \tau \in (0, \tau_0).$$

Take $t \in (0, \frac{\tau_0}{2})$. Then

$$R(B_t, \overline{M \setminus G}) \leq R(B_t, A_t) + R(A_t, \overline{M \setminus G}) \leq$$

$$\leq R(A_t, A_{2t}) + R(A_t, \overline{M \setminus G}) \leq$$

$$\leq 2R(A_t, \overline{M \setminus G}) + R(A_{2t}, \overline{M \setminus G}) < \epsilon.$$

This proves (4.6) .

To prove Theorem 4.1.3 take $\phi \in Z^*$, where Z^* is a residual subset of $Z(M)$ such that for any $x \in M$ (1.23) and (1.28) hold. Consider the map

$$\Phi_0 : (M^*, R_0) \to M^*$$

defined by $\Phi_0(G) = \omega(G)$. Let L be the set of continuity points of Φ_0.

Lemma 4.1.5. *For any $G_0 \in L$ the set $\omega(G_0)$ is Lyapunov stable.*

Proof. Let for $t \geq 0$ $K_t = \tilde{N}_t(G_0)$. It is easy to see that for $t \in [0, +\infty)$ the family K_t has the property (4.1). The same reasons as in the proof of Lemma 4.1.2 show that if $\omega(G_0)$ is not Lyapunov stable then there is $a > 0$ such that for any sequence $t_m > 0, \lim_{m\to\infty} t_m = 0$ we have

$$R(\omega(K_{t_m}), \omega(G_0)) \geq a. \tag{4.7}$$

It follows from lemma 4.1.4 that

$$\lim_{m\to\infty} R_0(K_{t_m}, G_0) = 0,$$

hence we obtain a contradiction between (4.7) and the choice of G_0. This proves the lemma.

We can represent L as

$$L = \bigcap_{k>0} L_k,$$

where L_k is the set of $\frac{1}{k}$-continuity points of Φ_0, that is the set of $G_0 \subset M^*$ having the property: there is a neighborhood U of G_0 in (M^*, R_0) such that for any $G_1, G_2 \in U$ we have $R(G_1, G_2) < \frac{1}{k}$. As any L_k is open it remains to show that any L_k is dense in (M^*, R_0). We are going to show that L is dense, hence L is residual in (M^*, R_0). This statement together with Lemma 4.1.5 will prove our theorem.

Fix $A \in M^*$ and define

$$F_A : (0, +\infty) \to M^*$$

by: $F_A(t) = \omega(\tilde{N}_t(A))$. Evidently F_A is an increasing map, so Lemma 4.1.3 implies that there is a residual subset C_A of $(0, +\infty)$ such that any $t \in C_A$ is a continuity point of F_A. Let

$$\tilde{L} = \{\tilde{N}_t(A) : A \in M^*, t \in C_A\}.$$

For fixed $A \in M^*$ the family $K_t = \tilde{N}_t(A)$ evidently satisfies condition (4.1) . As $\phi \in Z^*$ we can repeat the proof of Lemma 4.1.2 to show that for any $G \in \tilde{L}$ the set $\omega(G)$ is Lyapunov stable.

We claim that $\tilde{L} \subset L$. To prove this inclusion let us show that if for some $A \in M^*$ t_0 is a continuity point of F_A then $\tilde{N}_{t_0}(A)$ is a continuity point of Φ_0. Fix $\epsilon > 0$. As t_0 is a continuity point of F_A we can find $\delta > 0$ such that for any $t \in N_{2\delta}(t_0)$ we have

$$R(\omega(\tilde{N}_t(A)), \omega(\tilde{N}_{t_0}(A))) < \epsilon. \tag{4.8}$$

Take arbitrary $G \in M^*$ with

$$R_0(G, \tilde{N}_{t_0}(A)) < \delta. \tag{4.9}$$

As

$$G \subset \tilde{N}_{t_0+\delta}(A)$$

we obtain that

$$\omega(G) \subset \omega(\tilde{N}_{t_0+\delta}(A)).$$

It follows from (4.8) that

$$r(\omega(G), \omega(\tilde{N}_{t_0}(A)) < \epsilon. \tag{4.10}$$

It is evident that (4.9) implies

$$\overline{M \setminus G} \subset N_\delta(\overline{M \setminus \tilde{N}_{t_0}(A)}),$$

hence

$$M \setminus N_\delta(\overline{M \setminus \tilde{N}_{t_0}(A)}) \subset \overline{M \setminus (\overline{M \setminus G})} =$$
$$= \overline{\mathrm{Int} G} \subset G.$$

Therefore,

$$\omega(\tilde{N}_{t_0-\delta}(G)) \subset \omega(G). \tag{4.11}$$

It follows now from (4.8), (4.11) that

$$r(\omega(\tilde{N}_{t_0}(A)), \omega(G)) < \epsilon. \tag{4.12}$$

We obtain from (4.10), (4.12) that

$$R(\omega(G), \omega(\tilde{N}_{t_0}(A)) < \epsilon$$

so that $\tilde{N}_{t_0}(A)$ is a continuity point of Φ_0.

It remains now to prove that $\tilde{L}$ is dense in (M^*, R_0). Fix arbitrary $A \in M^*$. As C_A is a residual subset of $(0, +\infty)$ we can find a sequence $t_m \in C_A$ such that $\lim_{m\to\infty} t_m = 0$. We obtain from the definition of $\tilde{L}$ that any $\tilde{N}_{t_m}(A)$ is an element of $\tilde{L}$. It follows from Lemma 4.1.4 that

$$\lim_{m\to+\infty} R_0(\tilde{N}_{t_m}(A), A) = 0.$$

This proves that $\tilde{L}$ is dense.

V.Pogonysheva established also the following result [Po2].

Theorem 4.1.4. *For a generic system $\phi \in Z(M)$ there is a residual subset L of M^* such that for any $G \in L$ the set $\omega(G)$ is Lyapunov stable.*

We refer the reader to the original paper [Po2] for a proof of this theorem.

Now we investigate the process of "iterating of taking limit sets of neighborhoods". This process is obviously connected with the problem of Lyapunov stability of limit sets. Let us give exact definitions.

Let G be a subset of M. Fix a system $\phi \in Z(M)$ and $\delta > 0$. Denote

$$\omega_\delta^0(G) = \omega(G),$$

and for $k = 0, 1, \ldots$

$$\omega_\delta^{k+1}(G) = \omega(N_\delta(\omega_\delta^k(G))).$$

Theorem 4.1.5 [Pi7]. *A generic system $\phi \in Z(M)$ has the following property. Given $G \subset M$ and $\epsilon > 0$ there exists $\Delta > 0$ such that for $\delta \in (0, \Delta)$ and $k, l \geq 1$*

$$R(\omega_\delta^k(G), \omega_\delta^l(G)) < \epsilon.$$

Take a set $G \subset M$, a system $\phi \in Z(M)$, and define the set

$$\tilde{\omega}(G) = \overline{\bigcup_{x \in \overline{G}} R^\omega(x, \phi)}.$$

Lemma 4.1.6. *For any G, ϕ the set $\tilde{\omega}(G)$ is Lyapunov stable.*

Proof. To obtain a contradiction suppose that $\tilde{\omega}(G)$ is not Lyapunov stable. Then there exist sequences of points ξ_k, y_k and of numbers θ_k such that $y_k = \phi^{\theta_k}(\xi_k)$,

$$\lim_{k\to\infty} \xi_k = \xi \in \tilde{\omega}(G), \lim_{k\to\infty} y_k = y \notin \tilde{\omega}(G),$$

and $\lim_{k\to\infty} \theta_k = +\infty$. It follows from the definition of $\tilde{\omega}(G)$ that there exist points $x_k \in \overline{G}$, $\eta_k \in R^\omega(x_k, \phi)$ with

$$\lim_{k\to\infty} \eta_k = \xi.$$

Let x be a limit point of the sequence x_k. Then $x \in \overline{G}$ and it is easy to see that $y \in R^\omega(x, \phi)$, so that we obtain a contradiction which proves the lemma.

Lemma 4.1.7. *Given $G, \phi, \epsilon > 0$ there is a neighborhood $U(\epsilon)$ of the set $\omega(G)$ which has the following property: for any neighborhood U of $\omega(G)$ such that*

$$\omega(G) \subset U \subset U(\epsilon)$$

we have

$$r(\omega(U), \tilde{\omega}(G)) < \epsilon.$$

Proof. Fix G, ϕ, and $\epsilon > 0$. Apply previous lemma to find $\delta > 0$ such that for $x \in N_\delta(\tilde{\omega}(G))$ we have

$$d(\phi^k(x), \tilde{\omega}(G)) < \frac{\epsilon}{2}, k \geq 0.$$

We claim that always

$$\omega(G) \subset \tilde{\omega}(G). \tag{4.13}$$

To prove (4.13) take $y \in \omega(G)$ and find points $x_k \in G$, numbers t_k, $\lim_{k\to\infty} t_k = +\infty$ with $y = \lim_{k\to\infty} \phi^{t_k}(x_k)$. Let $x \in \overline{G}$ be a limit points of the sequence x_k. Then $y \in Q^\omega(x, \phi)$. As we always have $Q^\omega(x, \phi) \subset R^\omega(x, \phi)$ it follows that $y \in \tilde{\omega}(G)$.

It is evident that there is a neighborhood $U(\epsilon)$ of $\omega(G)$ such that $U(\epsilon) \subset N_\delta(\tilde{\omega}(G))$. Let us show that this neighborhood has the desired property. Take $U \subset U(\epsilon)$ and $z \in \omega(U)$. Fix points $\xi_k \in U$ and numbers $\lim_{k\to\infty} \tau_k = +\infty$ such that

$$\lim_{k\to\infty} \phi^{\tau_k}(\xi_k) = z,$$

it follows from the choice of δ that

$$d(\phi^{\tau_k}(\xi_k), \tilde{\omega}(G)) < \frac{\epsilon}{2}$$

therefore $d(z, \tilde{\omega}(G)) < \epsilon$. This proves the lemma.

Now consider again a residual subset Z^* of $Z(M)$ such that for any $\phi \in Z^*$ (1.23), (1.28) hold.

Lemma 4.1.8. *If $\phi \in Z^*, G \subset M$, and U is a neigborhood of $\omega(G)$ then $\tilde{\omega}(G) \subset \omega(U)$.*

Proof. Fix a point $z \in \tilde{\omega}(G)$. There exist finite δ_k - trajectories

$$(z_0^k, z_1^k, \ldots, z_{m(k)}^k)$$

with $z_0^k \in \overline{G}$, and a point $x_0 \in \overline{G}$ such that

$$\lim_{k\to\infty} z_0^k = x_0, \lim_{k\to\infty} z_{m(k)}^k = z,$$

$$\lim_{k\to\infty} m(k) = \infty, \lim_{k\to\infty} \delta_k = 0.$$

It is evident that $\omega_{x_0}(\phi)$, the ω - limit set of x_0 with respect to the system ϕ, is a subset of $\omega(G)$. Consider a neighborhood U of $\omega(G)$. As $\omega_{x_0}(\phi) \subset \omega(G)$ there is $m_0 > 0$ such that

$$y_0 = \phi^{m_0}(x_0) \in U.$$

As $\lim_{k\to\infty} z_{m_0}^k = y_0$, it is easy to see that

$$(y_0, z^k_{m_0+1}, \ldots, z^k_{m(k)-1}, z)$$

are finite δ'_k - trajectories for ϕ with $\lim_{k\to\infty} \delta'_k = 0$. Hence,

$$z \in R^\omega(y_0, \phi) = Q^\omega(y_0, \phi).$$

But as y_0 is an interior point of U this implies that $z \in \omega(U)$.

Let us now prove Theorem 4.1.5. Take a set $G \subset M$ and a system $\phi \in Z^*$. Analyzing the proof of Theorem 3.3.2 we can see that ϕ has the property described in this theorem. Therefore it follows from Lemma 4.1.6 that the set $\tilde{\omega}(G)$ is a quasi- attractor of ϕ, so that

$$\tilde{\omega}(G) = \bigcap_{m\geq 0} I_m,$$

where any I_m is an attractor, and $I_0 \supset I_1 \supset \ldots$.

Fix arbitrary $\epsilon > 0$. Find an attractor I_m (we call this attractor I below) such that

$$\tilde{\omega}(G) \subset I \subset N_\epsilon(\tilde{\omega}(G)).$$

There is a compact set K having the properties :

$$I \subset \mathrm{Int}K \subset K \subset D(I), K \subset N_\epsilon(\tilde{\omega}(G)).$$

Now take $\Delta > 0$ such that

$$\overline{N_\Delta(I)} \subset \mathrm{Int}K. \tag{4.14}$$

We claim that this Δ has the desired property. Let us use induction on k to show that for $k \geq 0$, $\delta \in (0, \Delta)$ we have

$$\omega^k_\delta(G) \subset I. \tag{4.15}$$

For $k = 0$ we have

$$\omega^0_\delta(G) = \omega(G) \subset \tilde{\omega}(G) \subset I$$

(see (4.13)). If (4.15) holds then

$$\omega^{k+1}_\delta(G) = \omega(N_\delta(\omega^k_\delta(G))) \subset \omega(N_\delta(I)).$$

We take here into account that for any sets $H_1 \subset H_2$ we have $\omega(H_1) \subset \omega(H_2)$. It follows from the choice of K, from (4.14), and from Lemma 3.3.7 that

$$\omega(N_\delta(I)) \subset \omega(\overline{N_\delta(I)}) \subset I.$$

This proves (4.15) for all $k \geq 0$.

Now take $k, l \geq 1$, let $k \geq l$. Apply Lemma 4.1.8 to establish inclusions

$$\tilde{\omega}(G) \subset \omega^l_\delta(G) \subset \omega^k_\delta(G) \subset N_\epsilon(\tilde{\omega}(G)).$$

It is evident that

$$R(\omega^k_\delta(G), \omega^l_\delta(G)) < \epsilon.$$

This proves the theorem.

4.2 Limit Sets for Diffeomorphisms Satisfying the STC

Let us consider a diffeomorphism ϕ which satisfies the STC.

Theorem 4.2.1. *Assume that K_t , $t \in (a,b)$, is a family of subsets of M which has the property (4.1). If ϕ is a diffeomorphism satisfying the STC then there exists a finite set C such that for $t \in (a,b)$ the set $\omega(K_t)$ is an attractor.*

To prove this result consider the decomposition

$$\Omega(\phi) = \Omega_1 \cup \ldots \cup \Omega_m$$

where Ω_i are basic sets. Let $\mu = \{1,\ldots,m\}$. For $\alpha \in \mu^*$ (that is for a subset of μ) let $\beta(\alpha)$ be the following subset of μ^* :

$$\beta(\alpha) = \alpha \cup \{j \in \mu : \text{ there is } i \in \alpha \text{ with } W^u(\Omega_i) \cap W^s(\Omega_j) \neq \emptyset\}.$$

Lemma 4.2.1. *Consider $\alpha \in \mu^*$. Then the set*

$$I = \bigcup_{i\in\alpha} \overline{W^u(\Omega_i)}$$

is an attractor of ϕ, and

$$D(I) = \bigcup_{j\in\beta(\alpha)} W^s(\Omega_j). \tag{4.16}$$

Proof. It follows from Lemma 0.4.2 that

$$W^u(\Omega_i) \cap W^s(\Omega_j) \neq \emptyset$$

is equivalent to

$$W^u(\Omega_j) \subset \overline{W^u(\Omega_i)}.$$

Therefore , I is an invariant compact set which coincides with

$$\bigcup_{j\in\beta(\alpha)} W^u(\Omega_j).$$

Let us show that I is Lyapunov stable. Introduce

$$\Delta = \min_{x\in\Omega_j, j\notin\beta(\alpha)} d(x,I).$$

Evidently, $\Delta > 0$. To obtain a contradiction suppose that I is not Lyapunov stable, hence there is $\varepsilon \in (0,\Delta/2)$, and sequences $\xi_k \in M, t_k > 0$ such that

$$\lim_{k\to\infty} d(\xi_k,I) = 0, d(\phi^{t_k}(\xi_k),I) \geq \varepsilon.$$

We can take numbers t_k so that

$$d(\phi^m(\xi_k), I) < \varepsilon \text{ for } 0 \leq m \leq t_k - 1. \tag{4.17}$$

Evidently, $\lim_{k\to\infty} t_k = +\infty$. Consider a limit point y of the sequence $\phi^{t_k}(\xi_k)$. There is a basic set Ω_j such that $y \in W^u(\Omega_j)$. If $j \in \beta(\alpha)$ then $y \in I$ which contradicts to the inequality $d(y, I) \geq \varepsilon$. If $j \notin \beta(\alpha)$ then there is $\tau < 0$ such that

$$d(\phi^\tau(y), \Omega_j) < \frac{\varepsilon}{2}.$$

Then for large k we have

$$d(\phi^{\tau+t_k}(\xi_k), \Omega_j) < \frac{\varepsilon}{2}, \tag{4.18}$$

and $0 < \tau + t_k < t_k$, so that (4.18) contradicts to (4.17). Therefore, I is Lyapunov stable.

Now apply Lemma 0.4.1 to find $\delta > 0$ having the followihg property : if $x \in W^s(\Omega_j), y \in W^u(\Omega_i)$, and $d(x, y) < \delta$ then

$$W^u(\Omega_i) \cap W^s(\Omega_j) \neq \emptyset.$$

It follows that if $x \in W^u(\Omega_i), i \in \beta(\alpha)$, then

$$N_\delta(x) \cap W^s(\Omega_j) = \emptyset$$

for $j \notin \beta(\alpha)$. Therefore any $x \in N_\delta(I)$ belongs to some $W^s(\Omega_j), j \in \beta(\alpha)$, so that

$$d(\phi^k(x), I) \leq d(\phi^k(x), \Omega_j),$$

hence

$$\lim_{k\to\infty} d(\phi^k(x), I) = 0.$$

This proves that I is an attractor, and establishes the formula (4.16) for $D(I)$.

Lemma 4.2.2. *Assume that G is a subset of M such that there exists $\alpha \in \mu^*$ with*

$$\overline{G} \subset \bigcup_{i\in\beta(\alpha)} W^s(\Omega_i);$$

$$(IntG) \cap W^s(\Omega_i) \neq \emptyset, i \in \alpha.$$

Then

$$\omega(G) = \bigcup_{i\in\alpha} \overline{W^u(\Omega_i)}.$$

Proof. It follows from Lemma 4.2.1 that the set

$$I = \bigcup_{i\in\alpha} \overline{W^u(\Omega_i)}$$

is an attractor. As $\overline{G}$ is a compact subset of $D(I)$ we obtain from Lemma 3.3.7 that $\omega(G) \subset I$.

Let us show that

$$I \subset \omega(G). \tag{4.19}$$

Consider a basic set Ω_i ,$i \in \alpha$, and fix a point $x \in (\text{Int}G) \cap W^s(\Omega_i)$. Apply Theorem 0.4.4 to find a point $p \in \Omega_i$ with $x \in W^s(p)$. As the set Int$G$ is open, and Per(ϕ) is dense in Ω_i we can find $p \in$Per(ϕ) such that

$$W^s(p) \cap \text{Int}G \neq \emptyset.$$

Take a small open n-dimensional ball $G_0 \subset G$ with $W^s(p) \cap G_0 \neq \emptyset$. It follows from Theorem 0.4.5 that

$$W^u(p) \subset \omega(G_0) \subset \omega(G).$$

As $\omega(G)$ is ϕ-invariant we obtain that

$$\overline{\bigcup_{r \in O(p,\phi)} W^u(r)} \subset \omega(G).$$

Now we can apply the statement (c) of Theorem 0.4.4 to show that for any $i \in \alpha$

$$\overline{W^u(\Omega_i)} \subset \omega(G).$$

This proves (4.19).

Now let us prove Theorem 4.2.1. Fix a basic set Ω_i of ϕ. Consider $t_i \in (a, b]$ which is defined in the following way. If there is $t \in (a, b)$ with

$$\overline{K_t} \cap W^s(\Omega_i) \neq \emptyset, W^s(\Omega_i) \cap \text{Int}K_t = \emptyset \tag{4.20}$$

then let t_i be equal to this t. Otherwise let $t_i = b$. It is evident that if $t_i \neq b$ then for $t < t_i$ we have

$$\overline{K_t} \cap W^s(\Omega_i) = \emptyset,$$

for $t > t_i$ we have

$$W^s(\Omega_i) \cap \text{Int}K_t \neq \emptyset.$$

Thus, we see that there is not more than one value $t \in (a, b)$ such that (4.20) holds. Let

$$C = (t_1, \dots, t_m),$$

here t_i correspond to basic sets Ω_i in the described way. Take $t \in (a, b) \setminus C$, and let

$$\alpha = \{i \in \mu : \overline{K_t} \cap W^s(\Omega_i) \neq \emptyset\}.$$

It follows that for any $i \in \alpha$ we have

$$W^s(\Omega_i) \cap \text{Int}K_t \neq \emptyset.$$

As

$$\overline{K_t} \subset \bigcup_{i \in \alpha} W^s(\Omega_i)$$

we obtain from Lemmas 4.1.1,4.1.2 that $\omega(G)$ is an attractor.

The following result is a corollary of Theorem 4.2.1.

Theorem 4.2.2 [Pi7]. *Assume that ϕ is a diffeomorphism satisfying the STC. Given $x \in M$ there exists a finite set $C(x)$ such that for any $r \in (0, +\infty) \setminus C(x)$ the set $\omega(N_r(x))$ is an attractor .*

Now take a set $G \subset M$, and define for a diffeomorphism ϕ which satisfies the STC, the sets $\omega_\delta^k(G)$ in the same way as it was done before Theorem 4.1.5. To investigate these sets we need the following simple statement.

Lemma 4.2.3 [Pi7]. *Let I be a compact invariant set for a diffeomorphism ϕ satisfying the STC. If the set I is Lyapunov stable then I is an attractor.*

Proof. To obtain a contradiction suppose that I is Lyapunov stable but I is not an attractor. It is shown in [Z] that in this case any neigborhood of I contains a compact invariant set K such that $K \cap I = \emptyset$.

Apply Lemma 0.4.1 to find $\delta > 0$ having the following property : if $\xi, \eta \in \Omega(\phi)$, and there are points $z \in W^s(\xi), y \in W^u(\eta)$ with $d(z, y) < \delta$, then

$$W^s(\xi) \cap W^u(\eta) \neq \emptyset.$$

Find a compact invariant set K for ϕ such that

$$K \cap I = \emptyset, K \subset N_\delta(I).$$

Any compact invariant set contains nonwandering points, take a point $\eta \in \Omega(\phi) \cap K$, and a neighborhood V of η such that

$$V \subset N_\delta(I), \overline{V} \cap I = \emptyset.$$

As $\mathrm{Per}(\phi)$ is dense in $\Omega(\phi)$ we can find a periodic point $\eta_0 \in V$. It follows from the choice of V that $O(\eta_0, \phi) \cap I = \emptyset$. Let ξ be a point of I such that $d(\eta_0, \xi) < \delta$. There is a point $z \in \Omega(\phi)$ with

$$\xi \in W^u(z).$$

As we have

$$\lim_{k \to -\infty} d(\phi^k(\xi), \phi^k(z)) = 0; \phi^k(\xi) \in I, k \in \mathbf{Z},$$

and as I is Lyapunov stable we obtain that $z \in I$. Therefore $W^u(z) \subset I$. Let Ω_i be the basic set which contains z. The set $W^u(z)$ is dense in Ω_i, hence, $\Omega_i \subset I$. As $\mathrm{Per}(\phi)$ is dense in Ω_i , we can find a periodic point $z_0 \in \Omega_i$ with

$$d(\eta_0, W^u(z_0)) < \delta.$$

Then

$$W^s(\eta_0) \cap W^u(z_0) \neq \emptyset,$$

it follows from Theorem 0.4.5 that

$$W^u(\eta_0) \subset \overline{W^u(z_0)} \subset I,$$

hence $\eta_0 \in I$. The contradiction we obtained proves the lemma.

Theorem 4.2.3 [Pi7]. *If ϕ is a diffeomorphism satisfying the STC then for any subset G of M there is $\Delta > 0$ such that if $\delta \in (0, \Delta), k > 1$ then*

$$\omega_\delta^k(G) = \omega_\delta^1(G).$$

Proof. If the set $\omega(G)$ is Lyapunov stable then it follows from Lemma 4.2.3 that $I = \omega(G)$ is an attractor. Find in this case $\Delta > 0$ such that

$$\overline{N_\Delta(I)} \subset D(I).$$

Then evidently for $\delta \in (0, \Delta), k \geq 1$

$$\omega_\delta^k(G) = \omega(G).$$

Consider now the case when the set $\omega(G)$ is not Lyapunov stable. Take a basic set Ω_i of ϕ. It is easy to understand that there is not more than one positive δ such that

$$\overline{N_\delta(\omega(G))} \cap W^s(\Omega_i) \neq \emptyset,$$
$$N_\delta(\omega(G)) \cap W^s(\Omega_i) = \emptyset.$$

Let this number be δ_i. Take $\Delta_1 \in (0, \min \delta_i)$. By the same reason as in the proof of Theorem 4.2.1 there are basic sets $\Omega_{i_1}, \dots, \Omega_{i_k}$ such that for any $\delta \in (0, \Delta_1)$ the set $\omega_\delta^1(G)$ concides with

$$\overline{W^u(\Omega_{i_1})} \cup \dots \cup \overline{W^u(\Omega_{i_k})},$$

so that $\omega_\delta^1(G)$ is an attractor. Now find $\Delta_2 > 0$ such that for $\delta \in (0, \Delta_2)$

$$\overline{N_\delta(\omega_\delta^1(G))} \subset D(\omega_\delta^1(G)).$$

Evidently, for $\delta \in (0, \Delta)$, where $\Delta = \min(\Delta_1, \Delta_2)$, we have

$$\omega_\delta^k(G) = \omega_\delta^1(G) \text{ for } k = 2, 3, \dots.$$

This proves the theorem.

V.Pogonysheva studied in [Po3] the dependence of the set $\omega(G, \phi)$ on both the set G and the system ϕ.

Theorem 4.2.4 [Po3]. *Assume that ϕ is a diffeomorphism satisfying the STC. Let G be an open subset of M such that for a set $\alpha = \{i_1, \ldots, i_k\}$ of indices for basic sets $\Omega_i, i \in \alpha$,we have*

$$\overline{G} \subset \bigcup_{i \in \alpha} W^s(\Omega_i); G \cap W^s(\Omega_i) \neq \emptyset, i \in \alpha. \tag{4.21}$$

Then given $\epsilon > 0$ there is $\delta > 0$ such that for any $\tilde{\phi} \in Z(M)$ with $\rho_0(\phi, \tilde{\phi}) < \delta$ and for any open subset $\tilde{G}$ of M with $R(G, \tilde{G}) < \delta$ we have

$$R(\omega(G, \phi), \omega(\tilde{G}, \tilde{\phi})) < \epsilon.$$

Proof. It follows from Lemmas 4.2.1, 4.2.2 that if G satisfies (4.21) then the set

$$I = \bigcup_{i \in \alpha} \overline{W^u(\Omega_i)} \tag{4.22}$$

is an attractor which coincides with $\omega(G, \phi)$.

Take arbitrary $\epsilon > 0$. As $\overline{G}$ is a compact subset of $D(I)$ (see Lemma 4.2.1) we can apply Lemma 3.3.7 to find a neighborhood K_0 of ϕ in $Z(M)$ such that for any system $\tilde{\phi} \in K_0$ the following is true : for arbitrary sequences $x_k \in G, t_k$ with $\lim_{k \to \infty} t_k = +\infty$ we have

$$\overline{\lim_{k \to \infty}} d(\tilde{\phi}^{t_k}(x_k), I) < \frac{\epsilon}{2}.$$

Obviously in this case

$$\omega(G, \tilde{\phi}) \subset \overline{N_{\frac{\epsilon}{2}}(I)} \subset N_\epsilon(\omega(G, \phi)). \tag{4.23}$$

Formula (4.22) shows that the attractor I coincides with the set $\omega(G', \phi)$ for any domain $G' \subset M$ such that the analogue of (4.21) (with G' instead of G) is true.

It is geometrically evident that we can find (if necessary decreasing ϵ) a domain G' which satisfies the analogue of (4.21) and such that

$$N_\epsilon(G') \subset G. \tag{4.24}$$

As ϕ satisfies the STC, ϕ is topologically stable (see Theorem 2.3.3), hence there is a neighborhood K_1 of ϕ such that for any $\tilde{\phi} \in K_1$ there is a continuous map h of M onto itself such that $h \circ \tilde{\phi} = \phi \circ h$, and $d(x, h(x)) < \epsilon/2$ for $x \in M$.

Fix a point

$$x \in \omega(G, \phi) = \omega(G', \phi).$$

Find sequences $x_k \in G', t_k$ with $\lim_{k \to \infty} t_k = +\infty$ such that

$$x = \lim_{k \to \infty} \phi^{t_k}(x_k).$$

Now take a system $\tilde{\phi} \in K_1$, and let h be a corresponding map. Take for any x_k a point $y_k \in h^{-1}(x_k)$, it follows from (4.24) that $y_k \in G$. Let y be a limit point of the sequence

$$\tilde{\phi}^{t_k}(y_k).$$

As

$$h(\tilde{\phi}^{t_k}(y_k)) = \phi^{t_k}(x_k)$$

we obtain that $d(x, y) \leq \epsilon/2$. Taking into account that $y \in \omega(G, \tilde{\phi})$ we see that

$$\omega(G, \phi) \subset N_\epsilon(\omega(G, \tilde{\phi})). \tag{4.25}$$

Now (4.23),(4.25) imply

$$R(\omega(G, \phi), \omega(G, \tilde{\phi})) < \epsilon. \tag{4.26}$$

It is easy to see that there is $\delta_1 > 0$ such that if $\tilde{G}$ is a domain with $R(G, \tilde{G}) < \delta_1$ then $\tilde{G}$ satisfies the analogue of (4.21), hence

$$\omega(\tilde{G}, \phi) = \omega(G, \phi).$$

The neighborhoods K_0, K_1 of ϕ depend really not on the domain G, but on $\epsilon > 0$, on our diffeomorphism ϕ, and on basic sets $\Omega_i, i \in \alpha$. Find $\delta_2 > 0$ such that

$$N_{\delta_2}(\phi) \subset K_0 \cap K_1.$$

Then if $\delta \in (0, \min(\delta_1, \delta_2))$, $R(G, \tilde{G}) < \delta$, and $\rho_0(\tilde{\phi}, \phi) < \delta$ we have

$$R(\omega(\tilde{G}, \tilde{\phi}), \omega(G, \phi)) = R(\omega(\tilde{G}, \tilde{\phi}), \omega(\tilde{G}, \phi)) < \epsilon$$

(we use here (4.26) with $\tilde{G}$ instead of G). This completes the proof.

Appendix A. Shadowing for Diffeomorphisms with Hyperbolic Structure

In this Appendix we give a proof of the following result.

Theorem A.1. *Let ϕ be a diffeomorphism of class C^1 which satisfies the STC. Then there exist constants $\delta^*, L^* > 0$ such that for any δ-trajectory $\xi = \{x_k : k \in \mathbf{Z}\}$ with $\delta < \delta^*$ there is a point x with*

$$d(\phi^k(x), x_k) \leq L^*\delta, k \in \mathbf{Z}.$$

This theorem is a generalization of the statement: if a diffeomorphism ϕ satisfies the STC then ϕ has the POTP. This last result was proved independently by C.Robinson [Robi2] (note that the proof in [Robi2] is not complete), by K. Sawada [Sa] (with a slighty different statement and in the case $\dim M \geq 2$), and by A.Morimoto [Mori2].

In the proof of Theorem A.1 below we fix a diffeomorphism ϕ of class C^1 which satisfies the STC. We do not mention the dependence of appearing constants on such characteristics of ϕ as

$$\max_{x \in M} \|D\phi(x)\|$$

and so on.

We begin by defining some objects. We say that K is a segment of $\mathbf{Z}$ if K is one of the following sets:

$$K = \mathbf{Z},$$

$$K = \{k : -\infty < k \leq k_1\},$$

$$K = \{k : k_1 \leq k \leq k_2\},$$

$$K = \{k : k_1 \leq k < +\infty\}.$$

If K is a segment of $\mathbf{Z}$, and $x \in M$ we say that $\{\phi^k(x) : k \in K\}$ is a *trajectory segment.*

Fix constants $C, \eta > 0$, $\lambda \in (0, 1)$. We say that a trajectory segment $I = \{\phi^k(x) : k \in K\}$ is a (C, λ, η)- *hyperbolic trajectory segment* (*h.t.s* below) if there exist linear subspaces $E^s_{x_k}, E^u_{x_k}$, $k \in K$, (here $x_k = \phi^k(x)$) such that:

(a) $D\phi(x_k)E^\sigma_{x_k} = E^\sigma_{x_{k+1}}, \sigma = s, u;$
for $k \in K$ with $k+1 \in K$;
(b) $E^s_{x_k} \oplus E^u_{x_k} = T_{x_k}M, k \in K$;
(c) $\angle(E^s_{x_k}, E^u_{x_k}) \geq \eta, k \in K$;
(d) if $v \in E^s_{x_k}$ then $|D\phi^m(x_k)v| \leq C\lambda^m|v|$
for $k \in K, m \geq 0$ such that $k+m \in K$;
if $v \in E^u_{x_k}$ then $|D\phi^m(x_k)v| \leq C\lambda^{-m}|v|$
for $k \in K, m \leq 0$ such that $k+m \in K$;

Of course, if I is a hyperbolic set with hyperbolicity constants C, λ then any trajectory of I is a (C, λ, η)-h.t.s., where $\eta = \eta(C, \lambda)$ (see [Pi8] for example).

One of basic objects of the proof is a *generalized hyperbolic set* ($g.h.s$ below). Fix again $C, \eta > 0, \lambda \in (0,1)$. We say that a subset I of M is a (C, λ, η)-g.h.s. if I is a set of one of the following types:

(a) I is a (C, λ, η)-h.t.s.;

(b) I is a hyperbolic set such that for any $x \in I$ the trajectory $O(x)$ is a (C, λ, η)-h.t.s.;

(c) I is a set of one of the forms:

$$I = O^+(x) \cup \tilde{I}, I = \tilde{I} \cup O^-(y), I = O^+(x) \cup \tilde{I} \cup O^-(y),$$

here $\tilde{I}$ is a hyperbolic set being a (C, λ, η)-g.h.s., $O^+(x), O^-(y)$ are (C, λ, η)-h.t.s., and

$$\lim_{k\to+\infty} d(\phi^k(x), \tilde{I}) = 0, \lim_{k\to-\infty} d(\phi^k(y), \tilde{I}) = 0.$$

It is well-known that trajectories in a neighborhood of a hyperbolic set are hyperbolic. The same is true for a neighborhood of a g.h.s. To be exact, the following statement holds (one can use the proofs of Theorem 2.1 in Chap.1 and of Theorems 1.2,1.3 in Chap.4 of [Pl1] to prove this statement).

Lemma A.1. *Let I be a $(\tilde{C}, \tilde{\lambda}, \tilde{\eta})$- g.h.s. Then given $C > \tilde{C}, \lambda \in (\tilde{\lambda}, 1), \eta < \tilde{\eta}$ there exist constants $\epsilon_1 = \epsilon_1(C, \lambda, \eta)$, $\Delta_1 = \Delta_1(C, \lambda, \eta)$ such that if $Y = \{y_k = \phi^k(y) : k \in K\}$ is a trajectory segment in $N_{\epsilon_1}(I)$ then Y is a (C, λ, η)-h.t.s with dim $E^s_{y_k}$ =dim $E^s_x, x \in I$, having the following property: for points y_k there exist discs $\tilde{W}^s(y_k), \tilde{W}^u(y_k)$ of diamΔ_1 such that:*

(a) $T_{y_k}\tilde{W}^\sigma(y_k) = E^\sigma_{y_k}, \sigma = s, u, k \in K$;

(b) $\phi(\tilde{W}^s(y_k)) \subset \tilde{W}^s(y_{k+1})$ for $k \in K$ with $k+1 \in K$;
$\phi^{-1}(\tilde{W}^u(y_k)) \subset \tilde{W}^u(y_{k-1})$ for $k \in K$ with $k-1 \in K$;

(c) if $z \in \tilde{W}^s(y_k)$ then

$$d(\phi^m(z), \phi^m(y_k)) \leq C\lambda^m d(z, y_k)$$

for $k \in K, m \geq 0$ such that $k+m \in K$;
if $z \in \tilde{W}^u(y_k)$ then

$$d(\phi^m(z), \phi^m(y_k)) \leq C\lambda^{-m} d(z, y_k)$$

for $k \in K, m \leq 0$ such that $k+m \in K$;

(d) if Y_1, Y_2 are two (C, λ, η)-h.t.s with $Y_i \subset N_{\epsilon_1}(I), i = 1, 2$, and if $y_1 \in Y_1, y_2 \in Y_2, d(y_1, y_2) < \epsilon_1$, then for any $x_1 \in \tilde{W}^s(y_1), x_2 \in \tilde{W}^u(y_2)$ we have

$$\angle(T_{x_1}\tilde{W}^s(y_1), T_{x_2}\tilde{W}^u(y_2)) \geq \eta.$$

Now we are going to show that if I is a g.h.s then there is a neighborhood U of I such that any δ-trajectory in U with small δ is ϵ-traced by a real trajectory with $\epsilon = L\delta$, and L depends only on hyperbolicity characteristics of I but not on δ. This statement (Lemma A.2) is a generalization of the ϵ-trajectory Theorem of D.Anosov [An2]. An analogous result was proved by R.Bowen [Bow]. Both Anosov and Bowen considered a hyperbolic set I and established the existence of a neighborhood U of I having the property: given $\epsilon > 0$ there is $\delta > 0$ such that any δ-trajectory in U is ϵ-traced by a real trajectory. The Lipschitz dependence of ϵ on δ stated in Lemma A.2 is easily obtained using techniques of [An2] or of [Bow]. We are following ideas of [Bow] in the proof below.

Lemma A.2. *Let I be a (C, λ, η)- g.h.s. There exist constants $\epsilon_0 = \epsilon_0(C, \lambda, \eta), \delta_0 = \delta_0(C, \lambda, \eta), L_0 = L_0(C, \lambda, \eta)$ having the following property: if K a segment of $\mathbf{Z}$, and $\xi = \{x_k\}$ is a δ-trajectory of ϕ with $\delta < \delta_0$ and such that*

$$\{x_k : k \in K\} \subset N_{\epsilon_0}(I)$$

then there is a point x with

$$d(\phi^k(x), x_k) \leq L_0\delta, k \in K.$$

Proof. Take $C_0 = 2C, \lambda_0 = \frac{1+\lambda}{2}, \eta_0 = \frac{\eta}{2}$, and apply Lemma A.1 to find $\epsilon_1(C_0, \lambda_0, \eta_0)$, $\Delta_1(C_0, \lambda_0, \eta_0)$, here C_0, λ_0, η_0 play the role of C, λ, η; and C, λ, η play the role of $\tilde{C}, \tilde{\lambda}, \tilde{\eta}$ in Lemma A.1.

It is geometrically evident that there exist constants $\gamma = \gamma(C, \lambda, \eta) > 1$, $\Delta_2 = \Delta_2(C, \lambda, \eta)$ which have the following property: if Y_1, Y_2 are two (C_0, λ_0, η_0)-h.t.s. with $Y_1, Y_2 \subset N_{\epsilon_1}(I)$, and $y \in Y_1, y_2 \in Y_2$, $d(y_1, y_2) < \Delta_2$, then there is a point

$$z \in \tilde{W}^s(y_1) \cap \tilde{W}^u(y_2)$$

with

$$d(z, y_i) \leq \gamma d(y_1, y_2), i = 1, 2.$$

Find a natural $m = m(C, \lambda, \eta)$ such that

$$2C_0\lambda_0^m\gamma < 1. \tag{.1}$$

Evidently there is $L_1 = L_1(m)$ with the property: if $\xi = \{x_k\}$ is a δ-trajectory of ϕ then

$$d(x_{k+j}, \phi^j(x_k)) \leq L_1\delta \text{ for } k \in \mathbf{Z}, 0 \leq j \leq m.$$

Find $\epsilon_2(C, \lambda, \eta)$ such that if $x \in N_{\epsilon_2}(I)$, and for $y \in I$ with $d(x, y) < \epsilon_2$ we have $O_0^m(y) \subset I$ then

$$O_0^m(x) \subset N_{\epsilon_1}(I).$$

Let us begin with the case of a finite δ-trajectory $\{x_0, \ldots, x_{rm}\}$ with natural r such that

$$\{x_0, \ldots, x_{rm}\} \subset N_{\frac{\epsilon_2}{2}}(I). \tag{.2}$$

Here we take $\delta < \delta_0$ where for $\delta_0 > 0$ we have

$$2L_1\gamma\delta_0 < \Delta_2, 2C_0L_1\gamma\delta_0 < \Delta_1, C_0\gamma L_1\delta_0 < \frac{\epsilon_2}{2}. \tag{.3}$$

Let us construct points $y_0, \ldots, y_k$ as follows. Let $y_0 = x_0$. It follows from (.2) that

$$O_0^m(x_0) \subset N_{\epsilon_1}(I),$$

hence $O_0^m(x)$ is a (C_0, λ_0, η_0)-h.t.s. Similarly, $O_0^m(x_m)$ is a (C_0, λ_0, η_0)-h.t.s. As

$$d(\phi^m(y_0), x_m) = d(\phi^m(x_0), x_m) \le L_1\delta \le 2L_1\gamma\delta_0 < \Delta_2$$

there is a point

$$y_1 \in \tilde{W}^u(\phi^m(y_0)) \cap \tilde{W}^s(x_m)$$

such that

$$d(y_1, \phi^m(y_0)) \le \gamma L_1\delta \le 2\gamma L_1\delta,$$

$$d(y_1, x_m) \le \gamma L_1\delta \le 2\gamma L_1\delta.$$

We use induction to construct points $y_k, 1 \le k \le r$, such that

$$y_k \in \tilde{W}^u(\phi^m(y_{k-1})) \cap \tilde{W}^s(x_{mk}), \tag{.4}$$

$$d(y_k, \phi^m(y_{k-1})) \le 2L_1\gamma\delta, \tag{.5}$$

$$d(y_k, x_{mk}) \le 2L_1\gamma\delta, \tag{.6}$$

Suppose that we obtained points $y_1, \ldots, y_k$ such that (.4)-(.6) hold. Then as $y_k \in \tilde{W}^s(x_{mk})$ we have

$$d(\phi^m(y_k), \phi^m(x_{km})) \le 2C_0\lambda_0^m\gamma L_1\delta \le L_1\delta,$$

as

$$d(\phi^m(x_{km}), x_{(k+1)m}) \le L_1\delta,$$

we obtain

$$d(\phi^m(y_k), x_{(k+1)m}) \le 2L_1\delta,$$

so that we can find

$$y_{k+1} \in \tilde{W}^u(\phi^m(y_k)) \cap \tilde{W}^s(x_{(k+1)m})$$

such that analogues of (.5),(.6) are true.

Take $x = \phi^{-rm}(y_r)$, and consider $s, 0 \le s \le r$. As

$$\phi^{sm}(x) = \phi^{sm-rm}(y_r)$$

we can write

$$d(\phi^{sm}(x), y_s) \leq d(\phi^{-m}(y_{s+1}), y_s)+$$

$$+d(\phi^{-2m}(y_{s+2}), \phi^{-m}(y_{s+1})) + \ldots + d(\phi^{(s-r)m}(y_r), \phi^{(s-r+1)}(y_{r-1})) =$$

$$\sum_{t=s+1}^{r} d(\phi^{(s-t)m}(y_t), \phi^{(s-t+1)m}(y_{t-1})) \leq$$

$$\leq 2\gamma L_1 \delta \sum_{t=s+1}^{r} C_0 \lambda_0^m.$$

Here we take into account (.4),(.5). Hence,

$$d(\phi^{sm}(x), y_m) \leq L_2 \delta,$$

where

$$L_2 = 2\gamma L_1 C_0 \frac{1}{1-\lambda_0^m}.$$

If Q is a Lipschitz constant for ϕ then

$$d(\phi^{sm+i}(x), \phi^i(y_s)) \leq Q^m L_2 \delta, 0 \leq i \leq m.$$

As $d(\phi^i(y_s), x_{sm+i}) \leq$

$$\leq d(\phi^i(y_s), \phi^i(x_{sm})) + d(\phi^i(x_{sm}), x_{sm+i}) \leq$$

$$\leq 2Q^m \gamma L_1 \delta + L_1 \delta, 0 \leq i \leq m,$$

finally we obtain that

$$d(\phi^k(x), x_k) \leq L_0 \delta, k \in \{0, \ldots, rm\},$$

where

$$L_0 = (2Q^m \gamma + 1)L_1 + Q^m L_2.$$

Let $\epsilon_0 = \frac{\epsilon_1}{2}$.

Now take arbitrary finite segment K of $\mathbf{Z}$, and a finite δ-trajectory $\{x_k : k \in K\} \subset N_{\epsilon_0}(I)$ with $\delta < \delta_0$. We can change indices so that

$$K = \{0 \leq k \leq k_1\},$$

find integer r_1, r with $k_1 + r_1 = rm$, $0 \leq r_1 \leq m-1$. It was shown that there is a point x with

$$d(\phi^k(x), x_k) \leq L_0 \delta, 0 \leq k \leq rm,$$

hence the desired inequality holds for $k \in K$.

For a δ-trajectory $\{x_k : k \in \mathbf{Z}\} \subset N_{\epsilon_0}(I)$ with $\delta < \delta_0$ we can find points x^r such that

$$d(\phi^k(x^r), x_k) \leq L_0 \delta, -rm \leq k \leq rm.$$

If x is a limit point of the sequence x^r as $r \to \infty$ then

$$d(\phi^k(x), x_k) \le L_0\delta, k \in \mathbf{Z}$$

(see the proof of Theorem 2.1.2 in the case dim$M \ge 2$). The cases $K = \{k_1 \le k \le +\infty\}, K = \{-\infty \le k \le k_1\}$ are treated analogously. This completes the proof.

Let us now state without a proof a result which is an analogue for diffeomorphisms of a theorem obtained by V.Pliss [Pl3] for periodic systems of differential equations. As previously we assume that ϕ satisfies the STC.

Lemma A.3. *There exist constants $\tilde{C}, \tilde{\eta}, \tilde{\beta} > 0$ and $\tilde{\lambda} \in (0,1)$ such that for any $x \in M$ and for any $T > 0$ we can find integers $k_0, \dots, k_m$ having the following properties:*

(a) the sets

$$O_{-\infty}^{k_0}(x), O_{k_0}^{k_1}(x), \dots, O_{k_m}^{+\infty}(x)$$

are $(\tilde{C}, \tilde{\lambda}, \tilde{\eta})$- h.t.s;

(b) $k_{i+1} - k_i \ge T, i = 0, \dots, m-1$;

(c) if $E_{i,0}^{\sigma}$ is $E_{\phi^{k_i}(x)}^{\sigma}$ in the h.t.s. $O_{k_{i-1}}^{k_i}(x)$, and if $E_{i,1}^{\sigma}$ is $E_{\phi^{k_i}}^{\sigma}(x)$ in the h.t.s $O_{k_i}^{k_i+1}(x), \sigma = s, u$, then

$$dim E_{i,0}^s < dim E_{i,1}^s, i = 0, \dots, m-1; \qquad (.7)$$

$E_{i,0}^u$ and $E_{i,1}^s$ are transversal, and

$$\angle(E_{i,0}^u, E_{i,1}^s) \ge \tilde{\beta}, i = 0, \dots, m-1. \qquad (.8)$$

Remark. If $O(x)$ is a trajectory of a hyperbolic set then we say below that the set $\{k_0, \dots, k_m\}$ is empty (formally we can take $k_0 = +\infty$). It follows immediately from (.7) that for any decomposition of a trajectory $O(x)$ into h.t.s described in Lemma A.3 we have $m \le n =$dim$M$.

Let us now apply techniques of K.Sawada [Sa] to describe limit sets of δ-trajectories (as $\delta \to 0$) for ϕ (see Lemmas A.4-A.6).

Lemma A.4. $CR(\phi) = \Omega(\phi)$.

Remark. This statement was proved in [Sa] in the case dim$M \ge 2$, and this was the reason for the restriction dim$M \ge 2$ in the main result of [Sa]. So in the case dim$M = 1$, that is for $M = S^1$, we give an independent proof.

Proof. If $M = S^1$ our diffeomorphism ϕ is a Morse-Smale diffeomorphism (see Sect. 0.4).

Denote by I_+ the union of stable periodics points of ϕ, and by I_- the union of stable periodic points of ϕ^{-1}. As dim$M = 1$, we have Per$(\phi) = I_+ \cup I_-$.

Evidently, I_+ is an attractor for ϕ. Fix arbitrary $\epsilon > 0$, and let $V_+ = N_\epsilon(I_+)$, $V_- = N_\epsilon(I_-)$. We may take ϵ so small that

$$\phi(\bar{V}_+) \subset V_+.$$

Hence there is $\Delta > 0$ such that for $x \in \phi(\bar{V}_+)$ we have

$$N_\Delta(x) \subset V_+.$$

As any trajectory of a Morse-Smale diffeomorphism tends to Per(ϕ) we see that the set

$$K = S^1 \setminus (V_+ \cup V_-)$$

is a compact subset of $D(I_+)$. Apply Lemma 3.1.2 to find m_0 such that for any $x \in K$ we have

$$\phi^{m_0}(x) \in \phi(V_+).$$

There exists $\Delta_0 > 0$ having the following property: for any δ-trajectory $\xi = \{x_k\}$ with $\delta < \Delta_0$ we have

$$x_{m_0} \in N_\Delta(\phi^{m_0}(x_0)).$$

Take $\delta < min(\Delta_0, \Delta)$. Consider a δ-trajectory $\xi = \{x_k\}$. Suppose that for some m we have $x_m \in V_+$. Then as

$$d(x_{m+1}, \phi(x_m)) < \delta < \Delta,$$

$$\phi(x_m) \in \phi(V_+) \subset \phi(\overline{V}_+)$$

we obtain that $x_{m+1} \in V_+$.

Now take a point

$$x \notin \overline{N_\epsilon(I_+ \cup I_-)}.$$

There exists a neighborhood W of x with

$$W \cap N_\epsilon(I_+ \cup I_-) = \emptyset.$$

If δ is small enough it follows from our previous considerations that for any δ-trajectory $\xi = \{x_k\}$ with $x_0 = x$ we have $x_m \notin W$ for $m \geq m_0$, hence $x \notin CR(\phi)$. We see that for arbitrarty $\epsilon > 0$

$$CR(\phi) \subset \overline{N_\epsilon(I_+ \cup I_-)} = \overline{N_\epsilon(Per(\phi))} = \overline{N_\epsilon(\Omega(\phi))},$$

hence

$$CR(\phi) \subset \Omega(\phi). \tag{.9}$$

It was shown in the proof of Theorem 1.4.1 that for any system ϕ we have

$$\Omega(\phi) \subset CR(\phi).$$

This proves our lemma in the case dim$M$ = 1. Now consider the case dim$M \geq 2$. It is sufficient again to show that for a diffeomorphism ϕ which satisfies the STC (.9) holds. Take a point $x \in CR(\phi)$. To obtain a contradiction suppose that $x \notin \Omega(\phi)$, and let $a = d(x, \Omega(\phi)) > 0$. It was shown in Lemma 2.2.7 that ϕ has a fine filtration, so by Theorem 1.4.2 ϕ has no C^0 Ω-explosions. Hence there is a heighborhood W of ϕ in $Z(M)$ such that for any system $\psi \in W$ we have

$$\Omega(\psi) \subset N_a(\Omega(\phi)),$$

so that $x \notin \Omega(\psi)$.

For any $\delta > 0$ we can find a δ-trajectory $\xi = \{x_k\}$ of ϕ and $m > 0$ with

$$x_0 = x, x_m = x.$$

Apply techniques of Lemma 1.2.1 to find a δ-trajectory $\tilde{\xi} = \{\tilde{x_k}\}$ where $\tilde{x}_k = x_k$ for $k \leq 0$, and for $k > m$, and the points $\tilde{x}_1, \ldots, \tilde{x}_m$ are distinct. Now the same reasons as in Lemma 1.2.2 (based on Lemma 0.3.3) show that there exists a system $\psi \in Z(M)$ with $\rho_0(\phi, \psi) < 2\delta$ and with $\psi^m(x) = x$. Then

$$x \in Per(\psi) \subset \Omega(\psi).$$

For δ small enough we have $\psi \in W$. The contradiction we obtained proves (.9).

Consider a sequence $\xi_m = \{x_k^m : k \in \mathbf{Z}\}$ of δ_m-trajectories for ϕ such that $\delta_m \to 0$ as $m \to \infty$, and let $\Xi_m = \bar{\xi}_m \in M^*$. As M^* is compact we can find a limit point Ξ of the sequence Ξ_m, suppose for simplicity that $\lim_{m\to\infty} \Xi_m = \Xi$.

Lemma A.5. *Let Ω_i be a basic set of ϕ. If the set*

$$B_i = (W^s(\Omega_i) \setminus \Omega_i) \cap \Xi \neq \emptyset$$

then there is a point y such that

$$B_i = O(y).$$

Proof. Take two distinct points $y, z \in B_i$, and find sequences $x_{k_1(m)}^m, x_{k_2(m)}^m \in \xi_m$ such that

$$y = \lim_{m\to\infty} x_{k_1(m)}^m, z = \lim_{m\to\infty} x_{k_2(m)}^m.$$

For an infinite number of m we have either

$$k_1(m) > k_2(m), \tag{.10}$$

or

$$k_2(m) > k_1(m). \tag{.11}$$

Consider the first case, then $y \in R(z, \phi)$, so there exists a sequence of $\frac{1}{m}$-trajectories $\{\eta_k^m : k \in \mathbf{Z}\}$ with

$$\eta_0^m = z, \eta_{k(m)}^m = y, k(m) \geq 1.$$

If the sequence $k(m)$ has a bounded subsequence, we may suppose that $k(m) \to \kappa$ as $m \to \infty$, then evidently $y = \phi^\kappa(z)$, so

$$y \in O^+(z).$$

To obtain a contradiction suppose that $k(m) \to \infty$ as $m \to \infty$. It is easy to understand that in this case for any $k \geq 0$ we have

$$y \in R(\phi^k(z), \phi).$$

Take a point $\tilde{z} \in \omega_z$, the ω-limit set of $O(z)$. Fix arbitrary $\delta > 0$. There is $\kappa > 0$ such that

$$\phi^\kappa(z) \in N_\delta(\tilde{z}).$$

Now for arbitrary $\delta_1 > 0$ find a δ_1-trajectory $\{\tilde{\eta}_k : k \in \mathbf{Z}\}$, and $\kappa_1 > \kappa$, such that

$$\tilde{\eta}_\kappa \in N_\delta(\tilde{z}), \tilde{\eta}_{\kappa_1} = y,$$

evidently,

$$y \in R(\tilde{z}, \phi). \tag{.12}$$

Take a point $\tilde{y} \in \omega_y$. As $y, z \in W^s(\Omega_i)$ we have $\tilde{y}, \tilde{z} \in \Omega_i$. The set Ω_i has a dense trajectory, hence

$$\tilde{z} \in R(\tilde{y}, \phi). \tag{.13}$$

As $\tilde{y} \in R(y, \phi)$,(.12) and (.13) imply that

$$y \in R(y, \phi),$$

so that $y \in CR(\phi)$. It follows from Lemma A.4 that $y \in \Omega(\phi)$. This contradicts to

$$y \in W^s(\Omega_i) \setminus \Omega_i.$$

The contradiction we obtained shows that if for an infinite number of m (.10) holds then $y \in O^+(z)$. Similarly one can show that if for an infinite number of m (.11) holds then $z \in O^+(y)$. This proves the lemma.

Now we can apply Lemma A.5 to describe the structure of a limit set Ξ of a sequence of δ_m-trajectories for $\delta_m \to 0$ as $m \to \infty$.

Lemma A.6. *For a set Ξ there exist basic sets $\Omega_{i_1}, \ldots, \Omega_{i_m}$ of ϕ, and points $z_1, \ldots, z_{m-1}$ such that*

$$\Xi \cap \Omega_{i_j} \neq \emptyset, j = 1, \ldots, m; \tag{.14}$$

$$z_j \in W^u(\Omega_{i_j}) \cap W^s(\Omega_{i_{j+1}}), j = 1, \ldots, m-1; \tag{.15}$$

$$\Xi \setminus \Omega(\phi) = \bigcup_{j=1}^{m-1} O(z_j). \tag{.16}$$

Proof. Take a basic set Ω_i of ϕ and denote

$$\Xi_i = (\Xi \setminus \Omega(\phi)) \cap W^s(\Omega_i).$$

As

$$M = \bigcup_i W^s(\Omega_i)$$

we obtain that

$$\Xi \setminus \Omega(\phi) = \bigcap_i \Xi_i.$$

By Lemma A.5, either $\Xi_i = \emptyset$ or there is a point z_i such that $\Xi_i = O(z_i)$. Find a basic set Ω_j such that $z_i \in W^u(\Omega_j)$. As the set Ξ is invariant and compact we obtain that

$$\Xi \cap \Omega_j \neq \emptyset.$$

It remains to change indices of Ω_i, z_j to obtain the statement of our lemma.

As M^* is compact the following statement is obviously true.

Lemma A.7. *Given $\epsilon > 0$ there exists $\delta(\epsilon)$ such that for any δ-trajectory ξ with $\delta < \delta(\epsilon)$ there is a set Ξ which has the properties described in Lemma A.6 and for which $R(\bar{\xi}, \Xi) < \epsilon$.*

Let $\tilde{C}, \tilde{\lambda}, \tilde{\eta}, \tilde{\beta}$ be the constants given for ϕ by Lemma A.3. We fix till the end of the proof of Theorem A.1 constants

$$C > \tilde{C}, \eta \in (0, \tilde{\eta}), \beta \in (0, \tilde{\beta}), \lambda \in (\tilde{\lambda}, 1).$$

Apply Lemma A.2 to find corresponding $\epsilon_1(C, \lambda, \eta), \Delta_1(C, \lambda, \eta)$. We can find $\tilde{\epsilon}_1(C, \lambda, \eta, \beta) \leq \epsilon_1, \tilde{\Delta}(C, \lambda, \eta, \beta) \leq \delta_1$ such that the following holds. If for a point $x \in M$

$$O^{k_i}_{k_{i-1}}(x), O^{k_{i+1}}_{k_i}(x)$$

are two $(\tilde{C}, \tilde{\lambda}, \tilde{\eta})$-h.t.s described in Lemma A.3, and if Y_1, Y_2 are two (C, λ, η)-h.t.s. with

(a) $Y_1 \subset N_{\tilde{\epsilon}_1}(O^{k_i}_{k_{i-1}}(x)), Y_2 \subset N_{\tilde{\epsilon}_1}(O^{k_{i+1}}_{k_i}(x))$;
(b) $z_j \in Y_j \cap N_{\tilde{\epsilon}_1}(\phi^{k_i}(x)), j = 1, 2$;
then for any $x \in \tilde{W}^u(z_1), y \in \tilde{W}^s(z_2)$ we have

$$\angle(T_x\tilde{W}^u(z_1), T_y\tilde{W}^s(z_2)) \geq \beta. \tag{.17}$$

For simplicity of notation we denote $\epsilon = \tilde{\epsilon}_1, \Delta = \tilde{\Delta}_1$ below.

Consider a (C, λ, η)-h.t.s. $Y = \{y_k = \phi^k(y) : k \in K\}$ which has properties described in Lemma A.1. Let $\sigma = \dim E^s_{y_k}$. Results on geometry of hyperbolic sets obtained on Chap.4 of [Pl1] show that we can introduce coordinates $(a, b), a \in \mathbf{R}^\sigma$, $b \in \mathbf{R}^{n-\sigma}$, in neighborhoods $U(y_k)$ of points $y_k \in Y$ as follows.

As

$$E^s_{y_k} \oplus E^u_{y_k} = T_{y_k}M$$

we can take $a \in E^s_{y_k}, b \in E^u_{y_k}$ as coordinates in $T_{y_k}M$. Let

$$B^\Delta_k = \{(a, b) \in T_{y_k}M : |a|, |b| \leq \Delta\}.$$

There exists $m_0 = m_0(C, \lambda, \eta)$ with the following property: for any $k \in K$ we can find a diffeomorphism

$$A_k : B_k^{\Delta} \to B_k^{\Delta}$$

such that

$$\rho_1(A_k, id) \leq m_0,$$

and the map

$$exp_{y_k} \circ A_k$$

defines coordinates (a, b) in $U(y_k) = exp_{y_k}(B_k^{\Delta})$ with properties described below. In these coordinates $\tilde{W}^s(y_k)$ is given by $b = 0$, $\tilde{W}^u(y_k)$ is given by $a = 0$. If for a point $(a, b) \in U(y_k)$ we have

$$\phi^l(a, b) = (a_l, b_l) \in U(y_{k+l})$$

for $-l_1 \leq l \leq l_2$ where $l_1, l_2 > 0$ then

$$|a_l| \leq C\lambda^l |a| \text{ if } 0 \leq l \leq l_2,$$

$$|b_l| \leq C\lambda^{-l} |b| \text{ if } -l_1 \leq l \leq 0.$$

Evidently there exist $m_i = m_i(C, \lambda, \eta), i = 1, 2$, such that if for $z_1, z_2 \in U(y_k)$ $|z_1 - z_2|$ is the Euclidean distance with respect to coordinates (a, b) then

$$m_1 d(z_1, z_2) \leq |z_1 - z_2| \leq m_2 d(z_1, z_2).$$

Below we say that $U(y_k)$ is the standard neighborhood of radius Δ centered at y_k.

Now fix a point z and a standard neighborhood $U(z)$ of radius Δ. Let (a, b) be coordinates in $U(z)$. Fix $N, l, g > 0$.

We say that D is an (N, l, g)-disc in $U(z)$ if D is a subset of $U(z)$ given by

$$a = \Phi(b)$$

where Φ is a map of class C^1 on $|b| \leq N$ such that $|\Phi(0)| \leq g$, and for b_1, b_2 with $|b_i| \leq N, i = 1, 2$, we have

$$|\Phi(b_1) - \Phi(b_2)| \leq l |b_1 - b_2|.$$

Below we fix Δ so small that if for two standard neighborhoods $U(z_1), U(z_2)$ we have $U(z_1) \cap U(z_2) \neq \emptyset$ then for any point $x \in U(z_1) \cap U(z_2)$ the map exp_x^{-1} is a diffeomorphism of $\overline{U(z_1) \cup U(z_2)}$ into $T_x M$, so that for any two discs D_1, D_2 in $U(z_1), U(z_2)$, and for points $x_i \in D_i, i = 1, 2$, we can define

$$\angle(T_{x_1} D_1, T_{x_2} D_2)$$

(see Sect.0.4).

Let Y_1, Y_2 be (C, λ, η)-h.t.s. Take $z_i \in Y_i, i = 1, 2$. The following statement is geometrically evident.

Lemma A.8. *Given Δ_1 there exist constants $N, l, l_0, \nu, g, \gamma_0$ (depending only on $C, \lambda, \eta, \beta, \Delta_1$) such that if*

(a) $dim\tilde{W}^u(z_1) \geq dim\tilde{W}^u(z_2)$;

(b) for any $x \in \tilde{W}^u(z_1), y \in \tilde{W}^s(z_2)$ (.17) holds;

(c) for some $\gamma \in (0, \gamma_0)$ $d(z_1, z_2) \leq \gamma\Delta_1$;

then

(a) $\tilde{W}^u(z_1), \tilde{W}^s(z_2)$ have a point w of transversal intersection with

$$d(w, z_i) \leq N\gamma, i = 1, 2;$$

(b) if D is an $(\nu\gamma N, l_0, \gamma g)$- disc in $U(z_1)$ then D contains a subset $\tilde{D}$ being a $(\gamma N, l, \gamma(g + 2\nu N))$-disc in $U(z_2)$.

Take $L_0 = L_0(C, \lambda, \eta), \delta_0 = \delta_0(C, \lambda, \eta)$ given by Lemma A.2, and let $\Delta_1 = 2L_0$. Apply Lemma A.8 to find corresponding $N, l, l_0, \nu, g, \gamma_0$. Fix $T > 0$ such that

$$C^2\lambda^{2T} l < l_0, \frac{\lambda^{-T}}{C} > \nu, \frac{2C\lambda^T \nu N}{1 - C\lambda^T} < g. \tag{.18}$$

Consider a limit set Ξ of δ-trajectories (as $\delta \to 0$) which has the structure described by Lemma A.6. Denote $H_i = \Xi \cap \Omega_i$ for basic sets Ω_i of ϕ. To simplify notation let us suppose that

$$\Xi \setminus \Omega(\phi) = \bigcup_{j=1}^{m-1} O(z_j),$$

and

$$z_j \in W^u(\Omega_j) \cap W^s(\Omega_{j+1}), j = 1, \ldots, m - 1.$$

For a point $z_j, j = 1, \ldots, m - 1$, consider a decomposition

$$O(z_j) = O_{-\infty}^{k_0}(z_j) \cup \ldots \cup O_{k_r}^{+\infty}(z_j)$$

given by Lemma A.3 and such that

$$k_{i+1} - k_i \geq T, i = 0, \ldots, r - 1.$$

Let $\kappa_i = \{k_0, \ldots, k_r\}$. Define two subsets Z_1, Z_2 of the set $\{z_1, \ldots, z_{m-1}\}$: $z_j \in Z_1$ if $\kappa_j = \emptyset$ (that is if $O(z_j)$ is a hyperbolic trajectory), otherwise $z_j \in Z_2$.

Let us pay more attention to the structure of the set Ξ in the case $Z_2 \neq \emptyset$. Let in this case

$$i_1 = min\{i \in \{0, \ldots, m - 1\} : z_i \in Z_2\},$$

$$\kappa_{i_1} = \{k_0, \ldots, k_p\},$$

$$i_2 = min\{i \in \{0, \ldots, m - 1\} : i > i_1, z_i \in Z_2\},$$

$$\kappa_{i_2} = \{\tilde{k}_0, \ldots, \tilde{k}_p\},$$

and so on (to simplify notation we omit i_1, i_2 in indices).

If $i_1 = 1$ let

$$\Lambda_1 = H_1 \cup O^{k_0}_{-\infty}(z_1),$$

if $i_1 > 1$ let

$$\Lambda_1 = H_1 \cup O(z_1) \cup H_2 \cup \ldots \cup H_{i_1} \cup O^{k_0}_{-\infty}(z_{i_1}).$$

After that let

$$\Lambda_2 = O^{k_1}_{k_0}(z_{i_1}), \ldots, \Lambda_{p+1} = O^{k_p}_{k_{p-1}}(z_{i_1}),$$

$$\Lambda_{p+2} = O^{+\infty}_{k_p}(z_{i_1}) \cup H_{i_1+1} \cup \ldots \cup H_{i_2} \cup O^{\tilde{k}_0}_{-\infty}(z_{i_2}),$$

and so on.

As a result we obtain the representation of Ξ in the form

$$\Xi = \Lambda_1 \cup \ldots \cup \Lambda_\mu \tag{.19}$$

where

$$\Lambda_1 = \tilde{\Lambda}_1 \cup O^-(\chi_1),$$
$$\Lambda_\mu = O^+(\chi_{\mu-1}) \cup \tilde{\Lambda}_\mu,$$

for $1 < j < \mu$ either

$$\Lambda_j = O^+(\chi_{j-1}) \cup \tilde{\Lambda}_j \cup O^-(\chi_j),$$

or

$$\Lambda_j = O^{l_{j-1}}_0(\chi_{j-1}), \chi_j = \phi^{l_{j-1}}(\chi_{j-1}),$$

and

$$l_{j-1} \geq T.$$

Here $\tilde{\Lambda}_1, \ldots, \tilde{\Lambda}_\mu$ are (C, λ)- hyperbolic sets, and $O^+(\chi_j), O^-(\chi_j), O^{l_{j-1}}_0(\chi_{j-1})$ are (C, λ, η)-h.t.s.

Lemma A.9. *There exist $\tau(C, \lambda, \eta), \Delta^*(C, \lambda, \eta)$ such that if $O^{k_2}_{k_1}(x_1), O^{k_2}_{k_1}(x_2)$ are (C, λ, η)- h.t.s., $k_2 - k_1 \geq \tau$, and*

$$d(\phi^k(x_1), \phi^k(x_2)) < \Delta^*, k_1 \leq k \leq k_2, \tag{.20}$$

then $dimE^s_{x_1}$ in $O^{k_2}_{k_1}(x_1)$ coincides with $dimE^s_{x_2}$ in $O^{k_2}_{k_1}(x_2)$.

Proof. Take $\tau_0 = \tau_0(C, \lambda, \eta)$ such that

$$C\lambda^{\tau_0} < \frac{1}{2C}(\frac{1+\lambda}{2})^{-\tau_0}, \tag{.21}$$

and let $\tau = 3\tau_0$. Apply Lemma A.1 to find $\Delta^* = \Delta^*(C, \lambda, \eta)$ such that if for a point x we have

$$d(\phi^k(x), \phi^k(x_1)) < \Delta^*, k_1 \leq k \leq k_2,$$

then $O_{k_1}^{k_2}(x)$ is a $(2C, \frac{1+\lambda}{2}, \frac{\eta}{2})$- h.t.s. with dim$E_x^s$ =dim$E_{x_1}^s$. Take x_2 such that (.20) holds, and let $\tilde{E}_{\phi^k(x_2)}^s$ be the corresponding linear subspaces of $T_{\phi^k(x_2)}$ for the $(2C, \frac{1+\lambda}{2}, \frac{\eta}{2})$-h.t.s. $O_{k_1}^{k_2}(x_2)$. We can find k_0 such that

$$k_1 + \tau_0 < k_0 < k_2 - \tau_0.$$

To obtain a contradiction suppose that

$$dim E_{\phi^{k_0}(x_2)}^s > dim \tilde{E}_{\phi^{k_0}(x_2)}^s$$

(the case

$$dim E_{\phi^{k_0}(x_2)}^s < dim \tilde{E}_{\phi^{k_0}(x_2)}^s$$

is treated analogously). In this case we can find

$$v \neq 0, v \in E_{\phi^{k_0}(x_2)}^s \cap \tilde{E}_{\phi^{k_0}(x_2)}^u.$$

Then

$$\frac{1}{2C}\left(\frac{1+\lambda}{2}\right)^{-m}|v| \leq |D\phi^m(\phi^{k_0}(x_2))v| \leq C\lambda^m|v|$$

for $0 \leq m \leq k_0 + \tau$. The contradiction with (.21) proves our lemma.

Below we consider $T > \tau$. It follows from (.7) that for $j - 1, \ldots, \mu - 1$ $O_{-\tau}^0(\chi_j), O_0^\tau(\chi_j)$ are (C, λ, η)- h.t.s with different dimE_x^s, hence we obtain from Lemma A.9 that the points

$$\chi_j \notin \Omega(\phi), j = 1, \ldots, \mu - 1$$

(note that for our diffeomorphism ϕ trajectories in $\Omega(\phi)$ are evidently (C, λ, η)-h.t.s.).

It follows from Lemma A.7 that there is a function $E(\delta) \to 0$ as $\delta \to 0$ such that for any δ-trajectory ξ there is a limit set Ξ with

$$R(\overline{\xi}, \Xi) < E(\delta). \tag{.22}$$

Lemma A.10. *There exists a function $E_1(\delta)$ such that if we have a set Ξ in the form (.19), if for some δ- trajectory $\xi = \{x_k : k \in \mathbf{Z}\}$ (.22) holds, and if*

$$d(x_{k_j}, \chi_j) < E(\delta), j = 1, \ldots, \mu - 1,$$

then

$$d(x_k, \Lambda_1) < E_1(\delta), k \leq k_1,$$

$$d(x_k, \Lambda_2) < E_1(\delta), k_1 \leq k \leq k_2,$$

$$\ldots$$

$$d(x_k, \Lambda_\mu) < E_1(\delta), k \geq k_{\mu-1}.$$

Proof. To obtain a contradiction suppose for definiteness that there is $a > 0$, a sequence of $\frac{1}{m}$-trajectories $\xi_m = \{x_k^m : k \in \mathbf{Z}\}$, and a sequence of sets

$$\Xi_m = \Lambda_1^m \cup \ldots \cup \Lambda_{\mu(m)}^m$$

such that

$$R(\bar{\xi}_m, \Xi_m) < E(\frac{1}{m}), \lim_{m\to\infty} d(x_{k_1}^m(m), \chi_1^m) = 0,$$

and

$$\sup_{k \le k_1(m)} d(x_k^m, \Lambda_1^m) > a.$$

Passing to a subsequence of ξ_m we can find a limit set

$$\Xi^* = \Lambda^* \cup \ldots \cup \Lambda_{\mu^*}^*$$

such that

$$R(\bar{\xi}_m, \Xi^*) \to 0 \text{ as } m \to \infty,$$
$$R(\Xi_m, \Xi^*) \to 0 \text{ as } m \to \infty.$$

Find $\tau(m) < k_1(m)$ such that

$$d(x_{\tau(m)}^m, \Lambda_1^m) \ge a, \tag{.23}$$

and let $\chi_1^*, \chi_2^* \in \Xi^*$ be limit points of $x_{k_1(m)}^m, x_{\tau(m)}^m$, we suppose that

$$\lim_{m\to\infty} x_{k_1(m)}^m = \chi_1^*, \lim_{m\to\infty} x_{\tau(m)}^m = \chi_2^*. \tag{.24}$$

Take the number τ given by Lemma A.9, let

$$s_1^m = dimE_x^s \text{ in } O_{-\tau}^0(\chi_1^m), s_2^m = dimE_x^s \text{ in } O_0^\tau(\chi_1^m).$$

Passing to a subsequence, we can consider the case $s_1^m \equiv s_1$, $s_2^m \equiv s_2$, and

$$s_1 < s_2. \tag{.25}$$

It follows from Lemma A.9 and from (.25) that

$$O_{-\tau}^0(\chi_1^*), O_0^\tau(\chi_2^*)$$

are (C, λ, η)-h.t.s. with different $\dim E_x^s$, hence the trajectory $O(\chi_1^*)$ is not hyperbolic, and

$$\chi_1^* \notin \Omega(\phi). \tag{.26}$$

Let $\chi_1^* \in \Lambda_{i_1}^*$ in the decomposition of Ξ^*. It is easy to understand that (.23) implies the existence of $a_1 > 0$ such that for large m we have

$$r(O_{-\tau}^\tau(x_{\tau(m)}^m), \Lambda_1^m) \ge a_1 > 0$$

(we remind that

$$r(A, B) = \sup_{x\in A} d(x, B)),$$

hence

$$O^{\tau}_{-\tau}(x^m_{\tau(m)}) \subset N_{E(\frac{1}{m})}(\Lambda^m_2 \cup \ldots \cup \Lambda^m_{\mu(m)}).$$

As

$$R(O^{\tau}_{-\tau}(x^m_{\tau(m)}), O^{\tau}_{-\tau}(\chi^*_2)) \to 0 \text{ as } m \to \infty$$

we obtain from Lemmas A.1,A.3, and A.9 that if $\chi^*_2 \in \Lambda^*_{i_2}$ then $i_2 > i_1$.

It follows from the structure of the sets Λ^*_j that for points $\chi^*_j \in \Lambda^*_{i_j}, j = 1, 2$, with $i_2 > i_1$ we have

$$\chi^*_2 \in R(\chi^*_1, \phi).$$

But (.24) implies that

$$\chi^*_1 \in R(\chi^*_2, \phi),$$

so that

$$\chi^*_1 \in CR(\phi).$$

The contradiction with Lemma A.4 and with (.26) proves our lemma.

Take now $\epsilon_1 > 0$ such that $\epsilon_1 < min(\epsilon, \epsilon_0)$, here ϵ_0 is given by Lemma A.2, and ϵ is fixed after Lemma A.7. Find $\delta_1 > 0$ with

$$L_0\delta_1 < \frac{\epsilon_1}{2}$$

(L_0 is given by Lemma A.2) and such that

$$E(\delta), E_1(\delta) < \frac{\epsilon_1}{2} \text{ for } \delta \in (0, \delta_1).$$

Consider a δ-trajectory $\xi = \{x_k\}$ with $\delta < \delta_1$, and find a set Ξ such that (.22) holds. Let us begin with the case of a set Ξ such that $Z_2 \neq \emptyset$ and Ξ is represented in the form (.19). It follows from Lemma A.10 that in this case

$$d(x_k, \Lambda_1) < \frac{\epsilon_1}{2}, k \le k_1,$$

$$d(x_k, \Lambda_2) < \frac{\epsilon_1}{2}, k_1 \le k \le k_2,$$

$$\ldots$$

$$d(x_k, \Lambda_\mu) < \frac{\epsilon_1}{2}, k \ge k_{\mu-1}.$$

As the sets $\Lambda_1, \ldots, \Lambda_\mu$ are (C, λ, η)-g.h.s. we can apply Lemma A.2 to find points $z^+_1, z^-_2, z^+_2, \ldots, z^-_{\mu-1}$ such that

$$d(\phi^k(z^+_1), x_{k_1+k}) \le L_0\delta < \frac{\epsilon_1}{2}, k \le 0;$$

$$\phi^{k_2-k_1}(z^-_2) = z^+_2;$$

$$d(\phi^k(z^-_2), x_{k_1+k}) \le L_0\delta < \frac{\epsilon_1}{2}, 0 \le k \le k_2 - k_1;$$

$$\ldots$$

$$d(\phi^k(z_{\mu-1}^-), x_{k_{\mu-1}+k}) \leq L_0\delta < \frac{\epsilon_1}{2}, k \geq 0.$$

Take $\delta_2 \in (0, min(\delta_1, \gamma_0))$ (γ_0 is given by Lemma A.8). Assume that $\delta \in (0, \delta_2)$. Consider standard neighborhoods $U(z_1^+)$ in the (C, λ, η)- h.t.s. $O^-(z_1^+)$, and $U(z_2^-)$ in the (C, λ, η)- h.t.s. $O_0^{k_2-k_1}(z_2^-)$.

Let in coordinates (a, b) of $U(z_1^+)$

$$D_1^0 = \{(0, b) : |b| \leq \nu N\delta\}$$

(here N, ν are given by Lemma A.8). Evidently, D_1^0 is a $(\nu N\delta, l_0, g\delta)$-disc in $U(z_1^+)$. It follows from Lemma A.8 (here δ plays the role of γ) that D_1^0 contains a subset D_1 being an $(N\delta, l, (g + 2\nu N)\delta)$-disc in $U(z_2^-)$. The point $w^1 = D_1 \cap \tilde{W}^s(z_2^-)$ evidently belongs to $\tilde{W}^u(z_1^-)$, and

$$max(d(w^1, z_1^+), d(w^1, z_2^-)) \leq N\delta.$$

We take into account here that

$$d(z_1^+, z_2^+) \leq 2L_0\delta = \Delta_1\delta.$$

Take a point $z \in D_1^0$, in coordinates of $U(z_1^+)$ $z = (0, z^u)$ with $|z^u| \leq m_2\nu N\delta$. If $z_k = \phi^k(z)$ for $k \leq 0$, and $z_k = (0, z_k^u)$ in $U(\phi^k(z_1^+))$, then

$$|z_k^u| \leq C\lambda^{-k}m_2\nu N\delta,$$

hence

$$d(x_{k_1+k}, \phi^k(z)) \leq d(x_{k_1+k}, \phi^k(z_1^+)) +$$
$$+d(\phi^k(z_1^+), \phi^k(z)) \leq L_1\delta,$$

where

$$L_1 = L_0 + \frac{m_2\nu CN}{m_1}.$$

Let $w^1 = (w_0^{1,s}, 0)$ in coordinates of $U(z_2^-)$, then

$$|w_0^{1,s}| \leq (g + 2\nu N)\delta.$$

If $\phi^k(w^1) = (w_k^{1,s}, 0)$ in coordinates of $U(\phi^k(z_2^-)), 0 \leq k \leq k_2 - k_1$, then

$$|w_k^{1,s}| \leq C\lambda^k(g + 2\nu N)\delta,$$

and it follows from (.18) that if $k \geq T$ then we have

$$|w_k^{1,s}| \leq C\lambda^T(g + 2\nu N)\delta \leq g. \tag{.27}$$

Take a point $v \in D_1$, let $\phi^k(v) = (v_k^s, v_k^u)$ in coordinates of $U(\phi^k(z_2^-))$, $0 \leq k \leq k_2 - k_1$. As D_1 is an $(N\delta, l, (g + 2\nu N)\delta)$-disc in $U(z_2^-)$ we obtain that

$$|v_0^s| \leq l|v_0^u| + |w_0^{1,s}| \leq (l\nu N + g + 2\nu N)\delta,$$

hence for $0 \le k \le k_2 - k_1$

$$|v_k^s| \le C\lambda^k(g + \nu N(2 + l))\delta. \tag{.28}$$

Take a point $\tilde{v} \in D_1$ with $\phi^k(\tilde{v}) = (\tilde{v}_k^s, \tilde{v}_k^u)$ in $U(\phi^k(z_2^-))$. For $0 \le k \le k_2 - k_1$ we have

$$|v_k^s - \tilde{v}_k^s| \le C\lambda^k|v_0^s - \tilde{v}_0^s|, |v_k^u - \tilde{v}_k^u| \ge \frac{\lambda^{-k}}{C}|v_0^u - \tilde{v}_0^u|.$$

As $|v_0^s - \tilde{v}_0^s| \le l|v_0^u - \tilde{v}_0^u|$ and (.18) holds we see that for $k \ge T$ we have

$$|v_k^s - \tilde{v}_k^s| \le C^2\lambda^{2T} l|v_k^u - \tilde{v}_k^u| \le l_0|v_k^u - \tilde{v}_k^u|. \tag{.29}$$

While $\phi^k(v) \in U(\phi^k(z_2^-)), k \ge 0$,

$$|v_k^u| \ge \frac{\lambda^{-k}}{C}|v_0^u|.$$

Hence, if we take $\delta_3 \in (0, \delta_2)$ such that $\nu N\delta_3 < \Delta$, and if $\delta < \delta_3$, then for $k \ge T$

$$|v_k^u| \ge \nu|v_0^u| \tag{.30}$$

(see (.18)). It follows from (.27), (.29), (.30) that $\phi^{k_2-k_1}(D_1)$ contains a subset D_2^0 being a $(\nu N\delta, l_0, g\delta)$-disc in coordinates of $U(z_2^+)$. As

$$d(z_2^+, z_3^-) \le 2L_0\delta,$$

D_2^0 contains a subset D_2 being a $(N\delta, l, (g+2\nu N)\delta)$-disc in coordinates of $U(z_3^-)$.

Take a point $z \in D_2^0$, let for $k_1 - k_2 \le k \le 0$ $\phi^k(z) = (z_k^s, z_k^u)$ in coordinates of $U(\phi^k(z_2^+))$. As D_2^0 is a $(\nu N\delta, l_0, g\delta)$-disc in $U(z_2^+)$ we have $|z_0^u| \le \nu N\delta$, hence

$$|z_k^u| \le C\lambda^k \nu N\delta \le C\nu N\delta, k_1 - k_2 \le k \le 0. \tag{.31}$$

For our point z there is a point $v \in D_1$ such that

$$z = \phi^{k_2-k_1}v,$$

so we obtain

$$|z_k^s| \le C\lambda^{k+k_2-k_1}|v_0^s| \le C(l\nu N + g + 2\nu N)\delta. \tag{.32}$$

Therefore, for $k_1 - k_2 \le k \le 0$, we have from (.31), (.32) that

$$d(\phi^k(z), \phi^k(z_2^+)) \le \frac{C}{m_1}(g + \nu N(l + 3)),$$

and

$$d(\phi^k(z), x_{k_2+k}) \le L_2\delta$$

where

$$L_2 = L_0 + \frac{C}{m_1}(g + \nu N(l + 3)).$$

Continuing this process we can construct discs $D_3, \ldots, D_{\mu-2}$ which are $(N\delta, l, (g+2\nu N)\delta)$-discs in $U(z_4^-), \ldots, U(z_{\mu-1}^-)$ and the following holds: for any $z \in U(z_{\mu-1}^-)$ and for $k \leq 0$

$$d(\phi^k(z), z_{k_{\mu-1}+k}) \leq max(L_1, L_2)\delta.$$

As $D_{\mu-2}$ is an $(N\delta, l, (g+2\nu N)\delta)$-disc in $U(z_{\mu-1}^-)$ there is a point

$$x \in D_{\mu-2} \cap \tilde{W}^s(z_{\mu-1}^-)$$

such that if $x = (x^s, 0)$ in $U(z_{\mu-1}^-)$ then $|x^s| \leq (g+2\nu N)\delta$. For $k \geq 0$ we have

$$d(\phi^k(x), \phi^k(z_{\mu-1}^-)) \leq \frac{C}{m_1}(g+2\nu N)\delta,$$

and

$$d(\phi^k(x), x_{k_{\mu-1}+k}) \leq (L_0 + \frac{C}{m_1}(g+2\nu N))\delta \leq$$
$$\leq L_2\delta, k \geq 0.$$

Changing indices of ξ so that $k_{\mu-1} = 0$ we see that if we take $L^* = max(L_1, L_2)$ then

$$d(\phi^k(x), x_k) \leq L^*\delta, k \in \mathbf{Z}.$$

It remains to take $\delta^* = \delta_3$.

Now consider the case of the limit set Ξ such that $Z_2 = \emptyset$. It is evident that in this case Ξ is a (C, λ, η)-g.h.s. As for the constants δ^*, L^* we defined we have

$$\delta^* < \delta_0(C, \lambda, \eta), L^* > L_0(C, \lambda, \eta)$$

it follows that in this case the statement of Theorem A.1 is also true. This completes the proof.

Now let us prove Theorem 1.2.1. Fix arbitrary $\epsilon > 0$ and consider a subset V_ϵ of CLD(M): $\phi \in V_\epsilon$ if there is a neighborhood W of ϕ in $Z(M)$ and a number $\delta > 0$ such that for any $\phi_1, \phi_2 \in W$ the following holds: if $\xi = \{x_k : k \in \mathbf{Z}\}$ is a δ-trajectory for ϕ_1 then there exists $x \in M$ with

$$d(x_k, \phi_2^k(x)) < \epsilon, k \in \mathbf{Z}.$$

It follows from the definition that any $\psi \in W \cap$CLD(M) belongs to V_ϵ (we may take $\delta > 0$ for ψ the same as for ϕ), hence V_ϵ is open in CLD(M) for any $\epsilon > 0$.

To prove density of V_ϵ fix $\epsilon > 0$ and take a diffeomorphism ϕ which satisfies the STC. We claim that $\phi \in V_\epsilon$ (hence V_ϵ is dense in CLD(M) by Theorem 2.4.1).

Apply Theorem A.1 to find $\delta > 0$ such that if $\xi = \{x_k\}$ is a 2δ-trajectory for ϕ then ξ is $\frac{\epsilon}{2}$- traced by a trajectory of ϕ.

By Theorem 2.2.3 ϕ is topologically stable, hence we may take δ so small that for any system $\psi \in N_\delta(\phi)$ there is a homeomorphism $h : M \to M$ such that $\psi \circ h = h \circ \phi$ and $d(x, h(x)) < \frac{\epsilon}{2}$ for $x \in M$. Let $W = N_\delta(\phi)$.

Take two systems $\phi_1, \phi_2 \in W$. If $\xi = \{x_k\}$ is a δ-trajectory for ϕ_1 then ξ is a 2δ-trajectory for ϕ. Find $x \in M$ such that

$$d(x_k, \phi^k(x)) < \frac{\epsilon}{2}, k \in \mathbf{Z}.$$

Consider a homeomorphism h such that $\phi_2 \circ h = h \circ \phi$, $d(x, h(x)) < \frac{\epsilon}{2}$ for any x. Let $y = h(x)$. Then

$$d(x_k, \phi_2^k(y)) \leq d(x_k, \phi^k(x)) + d(\phi^k(x), \phi_2^k(y)) < \epsilon, k \in \mathbf{Z},$$

as $\phi_2^k(y) = h(\phi^k(x))$. So, $\phi \in V_\epsilon$.

The set

$$V = \bigcap_{\epsilon>0} V_\epsilon$$

is a residual subset of CLD(M). If $\phi \in V$ then ϕ has the POTP: fix $\epsilon > 0$, find a corresponding $\delta > 0$, and take $\phi_1 = \phi_2 = \phi$. This completes the proof.

Appendix B. Structure of Attractors with the STC on the Boundary

In this Appendix we consider a diffeomorphism ϕ of class C^1 such that I is an attractor of ϕ, and ϕ satisfies the STC on J, the boundary of I. The main result of this Appendix is the following theorem mostly proved by the author in [Pi1,Pi2].

Let $\Omega(J) = J \cap \Omega(\phi)$.

Theorem B.1. *If ϕ satisfies the STC on J then :*

(a) J is an attractor of ϕ;

(b) the set $\Omega(J)$ is hyperbolic, and $Per(\phi)$ is dense in $\Omega(J)$;

(c) for any points $p, q \in \Omega(J)$ the manifolds $W^s(p), W^u(q)$ are transversal;

(d) for any point $x \in J$ there is a point $p \in \Omega(J)$ such that $x \in W^u(p)$;

(e) the set $\Omega(J)$ can be decomposed :

$$\Omega(J) = \Omega_1 \cup \ldots \Omega_m \tag{.1}$$

so that $\Omega_1, \ldots, \Omega_m$ are disjoint compact ϕ-invariant sets, and each of these sets contains a dense trajectory.

Proof. Let Ψ denote the closure of the union of sets $\alpha_x \cup \omega_x$ for $x \in J$ .The following statement is a corollary of results of [Ma2] (see also Chap.13 of the book [Pi8]).

Lemma B.1. *There exist constants $C_1 > 0$, $\lambda_1 \in (0,1)$ such that :*

(a) $\Psi \subset H(C_1, \lambda_1)$;

(b) if for some $\tilde{C}, \tilde{\lambda}$ for a point $x \in J$ we have $x \in H(\tilde{C}, \tilde{\lambda})$ then $x \in H(C_1, \lambda_1)$.

As it was mentioned in Sect.0.4, we write $x \in H(C, \lambda)$ if the trajectory $O(x)$ is a hyperbolic trajectory with hyperbolicity constants C, λ.

For two points x_1, x_2 belonging to hyperbolic trajectories we write $x_1 \to x_2$ if there is a point y of transversal intersection of $W^u(x_1), W^s(x_2)$ such that $y \neq x_1, y \neq x_2$.

As I is an attractor there exists a neighborhood V of I such that $\overline{V} \subset D(I)$, hence

$$\lim_{k\to\infty} d(\phi^k(x), I) = 0, x \in \overline{V}.$$

Find a neighborhood V_1 of I such that

$$\overline{V_1} \cup \phi^{-1}(\overline{V_1}) \subset V,$$

and let

$$V_0 = \overline{V} \setminus V_1.$$

It is evident that V_0 is a compact set having the property : for any $x \in D(I) \setminus I$

$$O(x) \cap V_0 \neq \emptyset.$$

It follows from description of properties of hyperbolic sets given in [Pl1] that we can find a neighborhood W of Ψ, and numbers $C, \Delta > 0$, $\lambda \in (0,1)$ such that :

(a) $W \subset V_1$;
(b) $C > C_1, \lambda \in (\lambda_1, 1)$;
(c) if $O(x) \subset W$ then $x \in H(C,\lambda)$;
(d) if $x_1, x_2 \in V_1$, $O(x_1), O(x_2) \subset H(C,\lambda)$, and

$$0 < d(x_1, x_2) < \Delta,$$

then $x_1 \rightleftarrows x_2$ (this means that $x_1 \to x_2$ and $x_2 \to x_1$);
(e) if $x \in M$, $O(y) \subset H(C,\lambda)$, and

$$d(\phi^k(x), \phi^k(y)) \leq \Delta \tag{.2}$$

for $k \geq 0$ ($k \leq 0$) then $x \in W^s(y)$ (respectively, $x \in W^u(y)$).

For a sequence $\{Q_k : k \geq 0\}$ of subsets of M let

$$\lim Q_k = \{\lim_{k\to\infty} x_k : x_k \in Q_k\}.$$

Denote by Π the set of periodic points x of ϕ such that $O(x) \subset W$. For a point p and for a sequence $x_k \in \Pi$ we write $x_k \Rightarrow p$ if

$$\lim_{k\to\infty} x_k = p \text{ and } \lim_{k\to\infty} r(O(x_k), J) = 0.$$

It is well-known that as $\Psi \subset H(C_1, \lambda_1)$ for any point $p \in \Psi$ we can find a sequence $x_k \in \Pi$ such that $x_k \Rightarrow p$ (see [Pl1]).

Take two points $p, q \in \Psi$. We write below $p * q$ if
- either $q \in O(p)$;
- or $q \notin O(p)$, and there is a point $q' \in O(q)$ such that for any sequences $y_k, x_k \in \Pi$ with $x_k \Rightarrow p, y_k \Rightarrow q'$ we have $x_k \rightleftarrows y_k$ for large k.

Lemma B.2. *The relation * is an eqivalence relation.*

Proof. It is evident that * is reflexive and symmetric. Let us show that * is transitive. Take $p, q, r \in \Psi$ with $p * q, q * r$. If there are two trajectories in the set $\{O(p), O(q), O(r)\}$ which coincide then it follows immediately from the definition of * that $p * r$.

Consider the case of disjoint $O(p), O(q), O(r)$. Fix arbitrary sequences $x_k, y_k, y'_k, z'_k \in \Pi$ such that

$$x_k \Rightarrow p, y'_k \Rightarrow q' \in O(q), y_k \Rightarrow q, x'_k \Rightarrow r' \in O(r),$$

and let for $k \geq k_1$ $x_k \rightleftharpoons y'_k$, for $k \geq k_2$ $y_k \rightleftharpoons z'_k$. Suppose that $q' = \phi^m(q)$, take $x'_k = \phi^{-m}(x_k)$. It is evident that $x'_k \Rightarrow p' = \phi^{-m}(p)$, and that for $k \geq k_1$ we have $x'_k \rightleftharpoons y_k$. It is well-known (see Theorem 6.4 in [Pi8]) that $x \rightarrow y, y \rightarrow z$ implies $x \rightarrow z$, so that for $k \geq k_0 = \max(k_1, k_2)$ we have $x'_k \rightleftharpoons z'_k$, hence $p * r$. This completes the proof.

Fix a point $p \in \Psi$, and let

$$\Lambda_p = \{q \in \Psi : p * q\}.$$

It is evident that the set Λ_p is ϕ-invariant.

Lemma B.3. *If $p, q \in \Psi$ and $d(p, q) < \Delta$ then $p * q$.*

Proof. If $q \in O(p)$ our statement is evident. If $q \notin O(p)$ consider arbitrary sequences $y_k, x_k \in \Pi$ with $x_k \Rightarrow p, y_k \Rightarrow q$. There is k_0 such that for $k \geq k_0$ we have $x_k \neq y_k$, and $d(x_k, y_k) < \Delta$. It follows from the choice of Δ that $x_k \rightleftharpoons y_k$

Corollary 1 *Any set Λ_p is compact.*

Proof. Take a sequence $q_m \in \Lambda_p$, and let $\lim_{m\to\infty} q_m = r$. As the set Ψ is closed we have $r \in \Psi$. It follows from Lemma B.3 that $q_m * r$ for large m, by Lemma B.2 $p * q_m, q_m * r$ imply $p * r$.

Corollary 2 *If $q \in \Psi$ and $d(q, \Lambda_p) < \Delta$ then $q \in \Lambda_p$.*

It is evident now that for two points $p, q \in \Psi$ the sets Λ_p, Λ_q either are disjoint, or coincide. Corollaries of Lemma B.3 show that the number of different Λ_p is finite. Denote these sets $\Psi_1, \ldots, \Psi_\mu$.

It is easy to see that any set $\Psi_i, i = 1, \ldots, \mu$, is either a trajectory of a periodic point of ϕ which is isolated in Π , or for any points $p, q \in \mathrm{Per}(\phi) \cap \Psi_i$ we have $p \rightleftharpoons q' \in O(q)$. Indeed, if there are sequences $y'_k, x_k \in \Pi$ such that $x_k \Rightarrow p, y'_k \Rightarrow q' \in O(q)$ then for large k we have

$$x_k \rightleftharpoons p, x_k \rightleftharpoons y'_k, y'_k \rightleftharpoons q',$$

hence $p \rightleftharpoons q'$. Below we consider sets Ψ_i of the second kind (proofs for Ψ_i being orbits of isolated periodic points are trivial).

Let us introduce analogues of stable and unstable manifolds for Ψ_i :

$$W^s(\Psi_i) = \{x \in M : \lim_{k\to\infty} d(\phi^k(x), \Psi_i) = 0\},$$

$$W^u(\Psi_i) = \{x \in M : \lim_{k\to-\infty} d(\phi^k(x), \Psi_i) = 0\}.$$

The following statement is an easy consequence of Lemma A.2.

Lemma B.4. *Let Λ be a hyperbolic set of ϕ. There is a neighborhood $U(\Lambda)$ which has the property : given $\epsilon > 0$ there exists $\delta > 0$ such that if for two points x_1, x_2 we have $d(x_1, x_2) < \delta$, and*

$$O^+(x_1) \subset U(\Lambda), O^-(x_2) \subset U(\Lambda)$$

then we can find a point x such that

$$d(\phi^k(x), \phi^k(x_1)) < \epsilon, k \geq 0, \tag{.3}$$

$$d(\phi^k(x), \phi^k(x_2)) < \epsilon, k \leq 0. \tag{.4}$$

Let us fix neighborhoods K_i of the sets $\Psi_i, i = 1, \ldots, \mu$, such that :
(a) $\overline{K_i} \subset W$, $r(\overline{K_i}, \Psi_i) < \Delta/2$, $i = 1, \ldots, \mu$;
(b) $\phi(\overline{K_i}) \cup \overline{K_i} \cap \overline{K_j} = \emptyset$ if $i \neq j$;
(c) $\overline{K_i} \subset U(\Psi_i)$ (here $U(\Psi_i)$ are neighborhoods given by Lemma B.4).
Below we call these neighborhoods K_i canonical.
Denote

$$K = K_1 \cup \ldots \cup K_\mu.$$

Consider hyperbolic periodic points $p_1, \ldots, p_m$ of ϕ. Suppose that $p_1 = p_m$, and that there exist points $x_1, \ldots, x_{m-1}$, being points of transversal intersection of pairs $W^u(p_i), W^s(q_{i+1})$, where $q_i \in O(p_i)$. In this case we say that the set

$$\Gamma = \overline{O(x_1)} \cup \ldots \cup \overline{O(x_{m-1})}$$

is a *transversal homoclinic contour.*

The following statement is well-known (see Chap.4 of [Pl1], for example).

Lemma B.5. *If Γ is a transversal homoclinic contour, then :*

(a) Γ is a hyperbolic set;

(b) given neighborhoods $U_1, \ldots, U_{m-1}$ of trajectories $O(p_1), \ldots, O(p_{m-1})$, neighborhoods $V_1, \ldots, V_{m-1}$ of points $x_1, \ldots, x_{m-1}$, and a neighborhood U of the set Γ there is a periodic point p such that

$$O(p) \subset U; O(p) \cap U_i \neq \emptyset, O(p) \cap V_i \neq \emptyset, i = 1, \ldots, m-1.$$

Lemma B.6. *Let for a point x_0 the set $\overline{O(x_0)}$ belongs to a canonical neighborhood K_i, and let X_- , X_+ be the sets $\alpha_{x_0}, \omega_{x_0}$, respectively. Assume that for two sequences σ_m, ρ_m of periodic points we have* *(a)* $O(\sigma_m), O(\rho_m) \subset K_i$;

(b) $\lim_{m\to\infty} d(\rho_m, X_-) = 0, \lim_{m\to\infty} d(\sigma_m, X_+) = 0$.

Then given a neighborhood V *of* x_0 *we can find a natural* m_0 *such that for* $m \geq m_0$ *the set* V *contains a point of transversal intersection of* $W^u(O(\rho_m))$, $W^s(O(\sigma_m))$.

Proof. Consider a sequence of points $\xi_k = \phi^{\tau_k}(x_0)$, and a point $x_- \in X_- \cap \lim \rho_m$ such that

$$\lim_{k\to\infty} \xi_k = x_-, \lim_{k\to\infty} \tau_k = -\infty$$

There exist $C_0, \delta_0 > 0$, $\lambda_0 \in (0,1)$ such that if $d(\xi_k, x_-) < \delta_0$ then there are points η_k of transversal intersection of $W^u(x_-)$, $W^s(\xi_k)$ having the following properties :

$$d(\phi^m(\eta_k), \phi^m(\xi_k)) \leq C_0\lambda_0^m d(\eta_k, \xi_k), m \geq 0, \tag{.5}$$

and

$$\lim_{k\to\infty} d(\xi_k, \eta_k) = 0.$$

Take $\epsilon > 0$ so small that $N_\epsilon(x_0) \subset V$. Find $k > 0$ such that

$$d(\xi_k, x_-) < \delta_0, C_0 d(\xi_k, \eta_k) < \epsilon,$$

let $\tau_k = -\theta, \eta_k = u_-$. Fix a smooth disc d_- in $W^u(x_-)$ which intersects $W^s(O(x_0))$ at u_- and such that

$$\phi^\theta(d_-) \subset V$$

(this is possible as (.5) holds).

Find a sequence θ_k such that

$$\lim_{k\to\infty} \theta_k = +\infty, \lim_{k\to\infty} \phi^{\theta_k}(x_0) = x_+,$$

where

$$x_+ \in X_+ \cap \lim \sigma_m.$$

Fix a small smooth disc d_+ in $W^u(x_+)$ which contains x_+. There is a sequence of discs d_k lying in $W^u(O(x_0))$ such that

$$\phi^{\theta_k}(x_0) \in d_k,$$

and there exists a sequence of immersions $\chi_k : d_+ \to d_k$ with the property :

$$\lim_{k\to\infty} \rho_1(\chi_k, \mathrm{id}) = 0 \text{ in } C^1(d_+, M)$$

(below we write $\lim_{k\to\infty} \rho_1(d_+, d_k) = 0$ in this case).

As $x_+ \in \lim \sigma_m$ we can find m_1 such that for $m \geq m_1$ $W^s(\sigma_m)$ are transversal to d_+ . Techniques which is analogous to the proof of the λ-Lemma in [Pa1] shows that there exists a sequence of discs a_k such that

$$a_k \subset \phi^{\theta_k+\theta}(d_-), \lim_{k\to\infty} \rho_1(a_k, d_k) = 0.$$

Hence, $\lim_{k\to\infty} \rho_1(d_+, a_k) = 0$. Therefore, we can find k_0 such that for $k \geq k_0, m \geq m_1$ the discs a_k have points of transversal intersection with $W^s(\sigma_m)$. In $W^u(\rho_m)$ there are discs Δ_m such that

$$\lim_{m\to\infty} \rho_1(d_-, \Delta_m) = 0,$$

hence, there exists $m_0 \geq m_1$ with the properties :

$$\phi^\theta(\Delta_m) \subset V$$

for $m \geq m_0$, and

$$\phi^{\theta_k+\theta}(\Delta_m)$$

have points of transversal intersection with $W^s(\sigma_m)$. This completes the proof.

Lemma B.7. *Assume that $x_0 \in W^\sigma(\Psi_i)$, $\sigma = s$ or $\sigma = u$. Given a neighborhood U_0 of Ψ_i there is a trajectory $O(x) \subset U_0$ such that*

$$x_0 \in W^\sigma(O(x)).$$

Proof. We give a proof for $\sigma = s$ (for $\sigma = u$ the proof is analogous). Consider $U_0 \subset K_i$. Fix $\epsilon \in (0, \Delta)$, and find a corresponding $\delta > 0$ (we take $\Lambda = \Psi_i$ in Lemma B.4) so that

$$N_{2\epsilon+\delta}(\Psi_i) \subset U_0.$$

There is θ such that for $k \geq \theta$ we have

$$d(\phi^k(x_0), \Psi_i) < \delta.$$

Take $x_2 = \phi^\theta(x_0)$, and find $x_1 \in \Psi_i$ such that $d(x_1, x_2) < \delta$. By Lemma B.4 there is a point x for which (.3),(.4) hold. The inequality

$$r(O(x), \Psi_i) \leq \epsilon + \delta < 2\epsilon + \delta$$

implies that $O(x) \subset U_0$. It follows from (.4) that $x_2 \in W^s(x)$, hence $x_0 \in W^s(O(x))$.

Denote $M_0 = M \setminus I$. We say below that a set Ψ_i has the property B if there is a hyperbolic periodic point q of ϕ such that

$$O(q) \cap K_i \neq \emptyset, W^s(q) \cap M_0 \neq \emptyset. \tag{.6}$$

As usually we say that a set Ψ_i is locally maximal if there is a neighborhood U_i of Ψ_i with the following property : if for $x \in M$ we have $O(x) \subset U_i$ then $O(x) \subset \Psi_i$.

Lemma B.8. *Suppose that a set Ψ_i has the property B. Then :*

(a) Ψ_i *is locally maximal;*
(b) $W^\sigma(\Psi_i) = \bigcup_{p \in \Psi_i} W^\sigma(p), \sigma = s, u;$
(c) $W^u(\Psi_i) \subset J.$

Proof. Take a periodic point q such that (.6) holds. It is evident that $q \in I$. For a point

$$x \in W^s(q) \cap M_0$$

there is $r > 0$ such that $N_r(x) \subset M_0$. As $N_r(x)$ is an n-dimensional ball it is transversal to $W^s(q)$. It follows from Theorem 0.4.5 that

$$W^u(O(q)) \subset \overline{O(N_r(x))} \subset \overline{M_0}, \tag{.7}$$

hence $q \in J$. As $q \in \Psi$ we obtain from the choice of K_i that $q \in \Psi_i$.

Now let us show that if π is a periodic point of ϕ such that $\pi \in K_i$, and $O(\pi) \subset H(C, \lambda)$ then

$$\pi \in J, W^u(\pi) \subset J. \tag{.8}$$

Find $x \in \Psi_i$ such that $d(x, \pi) < \Delta/2$, and consider a sequence $x_k \in \Pi$ with $x_k \Rightarrow x$. For large k we have $d(x_k, \pi) < \Delta/2$ and $O(x_k) \subset H(C, \lambda)$, hence $x_k \rightleftharpoons \pi$. By the definition of Ψ_i we have $x_k \rightleftharpoons q'$ for some $q' \in O(q)$. As $q' \to x_k \to \pi$ we can apply the λ-Lemma to show that

$$W^u(O(\pi)) \subset \overline{W^u(O(q'))} \subset J,$$

evidently this proves (.8).

Take arbitrary point $x \in \Psi_i$, and let $x_k \in \Pi$, $x_k \Rightarrow x$. For large k we have $x_k \in K_i$, $O(x_k) \subset H(C, \lambda)$, hence $W^u(x_k) \subset J$. As unstable manifolds depend continuously on the trajectory (see [Pl1]), for a point $\xi \in W^u(x)$ we can find a sequence $\xi_k \in W^u(x_k)$ such that $\lim_{k\to\infty} \xi_k = \xi$. This implies that $\xi \in J$, so we proved the third statement of our lemma.

Now we are going to show that Ψ_i is locally maximal. Take a neighborhood U of Ψ_i such that $\overline{U} \subset K_i$. Suppose that for a point x_0 we have $\overline{O(x_0)} \subset U$, and denote by X_-, X_+ the α-limit set of $O(x_0)$ and the ω- limit set of $O(x_0)$, respectively. It follows from the choice of K_i that

$$r(X_-, \Psi_i) < \frac{\delta}{2}, r(X_+, \Psi_i) < \frac{\delta}{2}.$$

As the sets X_-, X_+ are hyperbolic we can find sequences of periodic points ρ_m, σ_m such that $\rho_m, \sigma_m \in K_i$ and

$$\lim \rho_m \subset X_-, \lim \sigma_m \subset X_+.$$

Fix a neighborhood V of x_0 and a number m_0 such that for $m \geq m_0$ V contains a point of transversal intersection of $W^u(O(\rho_m)), W^s(O(\sigma_m))$ (see Lemma B.6). It follows from our previous considerations that for large m we have $\rho_m, \sigma_m \in \Psi_i$, hence there exist points

$$\sigma'_m \in O(\sigma_m), y_m \in W^s(\rho_m) \cap W^u(\sigma'_m).$$

Note that as $W^u(\sigma'_m) \subset J$, and ϕ satisfies the STC on J , the points y_m are points of transversal intersection of $W^s(\rho_m), W^u(\sigma'_m)$. Let x_m be points of transversal intersection of $W^u(O(\rho_m)), W^s(O(\sigma_m))$ belonging to V, then

$$\Gamma_m = \overline{O(x_m)} \cup \overline{O(y_m)}$$

is a transversal homoclinic contour if m is large enough. It is evident now that $\Gamma_m \subset J$. It follows from Lemmas B.1,B.5 that $\Gamma_m \subset H(C, \lambda)$, hence there is a sequence of periodic points τ_m such that

$$\tau_m \in V; O(\tau_m) \subset H(C, \lambda); O(\tau_m) \cap K_i \neq \emptyset.$$

For large m we have $\tau_m \in \Psi_i$. As V is arbitrary this implies that $x_0 \in \Psi_i$. So, we proved that Ψ_i is locally maximal.

Lemma B.7 and the local maximality of Ψ_i imply the second statement of our lemma. This completes the proof.

Lemma B.9. *Assume that for sets* $\Psi^{(1)}, \ldots, \Psi^{(s)}$ *(from the collection* $\Psi_1, \ldots, \Psi_\mu$*) there are points*

$$x_i \in (W^u(\Psi^{(i)}) \cap W^s(\Psi^{(i+1)})) \setminus \Psi, i = 1, \ldots, s-1,$$

and that $\Psi^{(1)}, \ldots, \Psi^{(s)}$ *have the property B. Then* $\Psi^{(i)} \neq \Psi(j)$ *for* $1 \leq i < j \leq s$.

Proof. To obtain a contradiction suppose that two sets in the collection $\Psi^{(1)}, \ldots, \Psi^{(s)}$ coincide, let $\Psi^{(1)} = \Psi^{(s)}$. Apply Lemma B.8 to find points $\pi_i \in \Psi^{(i)}$, $\rho_{i+1} \in \Psi^{(i+1)}$ such that

$$x_i \in W^u(\pi_i) \cap W^s(\rho_{i+1}),$$

and to show that $x_i \in J$. The trajectories π_i, ρ_{i+1} are hyperbolic, the STC on J implies that any point x_i is a point of transversal intersection of $W^u(\pi_i)$, $W^s(\rho_{i+1})$. Choose neighborhoods V_i of points x_i so that $V_i \cap \Psi = \emptyset$. We can find periodic points p_i, q_{i+1} in arbitrary neighborhoods of $\Psi^{(i)}, \Psi^{(i+1)}$ (and hence in $\Psi^{(i)}, \Psi^{(i+1)}$) , and points $\xi_i \in V_i$ being points of transversal intersection of $W^u(p_i), W^s(q_{i+1})$. It follows from the definition of Ψ_i that either $p_i \in O(r_i)$ or there is a point $r'_i \in O(r_i)$ with $r'_i \to p_i$, hence there is a transversal homoclinic contour $\Gamma \subset J$ which contains $\xi_1, \ldots, \xi_{s-1}$.

The same reasons as in the proof of Lemma B.8 show that we can find a sequence of periodic points $\tau_m \in H(C, \lambda)$ such that

$$O(\tau_m) \cap V_i \neq \emptyset, O(\tau_m) \cap K^{(i)} \neq \emptyset, i = 1, \ldots, s-1.$$

For large m for a point $p \in \Psi^{(1)}$ and for some points $\tau'_m \in O(\tau_m)$ we have $p \to \tau'_m$, hence $\tau_m \in J$, and $\tau_m \in \Psi$. The contradiction with the choice of neighborhoods V_i proves our lemma.

Lemma B.10. *Assume that for two sets $\Psi^{(1)}, \Psi^{(2)}$ (from the collection $\Psi_1, \dots, \Psi_\mu$) we have :*

(a) $\Psi^{(1)}$ has the property B;

(b) $W^u(\Psi^{(1)}) \cap W^s(\Psi^{(2)}) \neq \emptyset$.

Then $\Psi^{(2)}$ has the property B.

Proof. Fix a point

$$x_0 \in W^u(\Psi^{(1)}) \cap W^s(\Psi^{(2)}).$$

By Lemma B.8 we can find a point $\gamma^{(1)} \in \Psi^{(1)}$ such that $x_0 \in W^u(\gamma^{(1)})$. Apply Lemma B.7 to find a point $\gamma^{(2)}$ with

$$x_0 \in W^s(\gamma^{(2)}), \overline{O(\gamma^{(2)})} \subset K^{(2)}.$$

As $x_0 \in J$, $\gamma^{(1)}$ and $\gamma^{(2)}$ are hyperbolic, and the STC holds on J , we see that $W^u(\gamma^{(1)}), W^s(\gamma^{(2)})$ are transversal at x_0. Fix a point x_+ belonging to X_+, the ω-limit set of $O(\gamma^{(2)})$. The set X_+ is a subset of $K^{(2)}$, hence X_+ is hyperbolic. Fix a sequence of hyperbolic periodic points τ_m such that $x_+ \in \lim \tau_m$. Fix also a sequence $\pi_m \in \Pi$ with $\pi_m \Rightarrow \gamma^{(1)}$. It was shown in the proof of Lemma B.8 that there is a periodic point $p \in \Psi^{(1)}$ such that $W^s(p) \cap M_0 \neq \emptyset$ and $p \to \pi_m$ for large m . As $W^u(\gamma^{(1)}), W^s(\gamma^{(2)})$ are transversal at x_0 we can find $\pi'_m \in O(\pi_m)$ such that for large m we have $\pi'_m \to \tau_m$, hence

$$W^s(\tau_m) \cap M_0 \neq \emptyset.$$

This implies that $\tau_m \in J$ for large m, so that $\Psi^{(2)}$ has the property B.

Lemma B.11. *Assume that for a sequence of points ξ_k , for a sequence of positive numbers α_k , and for two distinct sets $\Psi^{(i)}, \Psi^{(j)}$ in the collection $\Psi_1, \dots, \Psi_\mu$ we have*

$$\lim_{k\to\infty} d(\xi_k, \Psi^{(i)}) = 0, \lim_{k\to\infty} d(\phi^{\alpha_k}(\xi_k), \Psi^{(j)}) = 0,$$

$$\lim_{k\to\infty} r(O_0^{\alpha_k}(\xi_k), J) = 0.$$

Then we can find points $z_1, \dots, z_\sigma \in \lim O_0^{\alpha_k}(\xi_k)$ and sets $\Psi^{(i_1)} = \Psi^{(i)}$, $\Psi^{(i_2)}$, ... , $\Psi^{(i_\sigma)} = \Psi^{(j)}$ such that

$$z_s \in W^u(\Psi^{(i_s)}) \cap W^s(\Psi^{(i_{s+1})}), s = 1, \dots, \sigma - 1.$$

Proof. Let

$$X = \lim O_0^{\alpha_k}(\xi_k),$$

evidently $X \subset J$. As the sets $\Psi^{(i)}, \Psi^{(j)}$ are distinct, invariant and compact we obtain that $\lim_{k\to\infty} \alpha_k = +\infty$. Consider the set of all sequences of segments

$$\sigma_k = [\theta_k^-, \theta_k^+] \subset [0, \alpha_k]$$

such that

(a) $\{\phi^t(\xi_k) : \theta_k^- \leq t \leq \theta_k^+\} \subset K^{(i)}$;
(b) $\lim_{k\to\infty} (\theta_k^+ - \theta_k^-) = +\infty$.
It is evident that this set is non-empty. Let for any k

$$\tau_k^{(1)} = \max \theta_k^+$$

(here we take maximum for all segments σ_k with fixed k). It follows from our construction that

$$\phi^{\tau_k^{(1)}}(\xi_k) \in K^{(i)}, \eta_k^{(1)} = \phi^{\tau_k^{(1)}+1}(\xi_k) \notin K^{(i)},$$

hence

$$\eta_k^{(1)} \in \overline{\phi(K^{(i)})} \setminus K^{(i)}.$$

Let z_1 be a limit point of the sequence $\eta_k^{(1)}$, we suppose that

$$\lim_{k\to\infty} \eta_k^{(1)} = z_1.$$

We claim that

$$z_1 \in X \cap (W^u(\Psi^{(i)}) \setminus W^s(\Psi^{(i)})). \tag{.9}$$

It is evident that $z_1 \in X \subset J$ and that for $t \leq -1$ we have

$$\phi^t(z_1) \in K^{(i)}.$$

Hence X_- , the α-limit set of $O(z_1)$ is a subset of

$$\Psi^{(i)} = \Psi \cap K^{(i)}.$$

So, we obtain that $z_1 \in W^u(\Psi^{(i)})$. To obtain a contradiction suppose that $z_1 \in W^s(\Psi^{(i)})$. There exist numbers $\tau_0, \Delta_0 > 0$ such that

$$d(\phi^t(z_1), \Psi^{(j)}) \geq 2\Delta_0 \text{ for } t \geq 0,$$

$$\phi^t(z_1) \in K^{(i)} \text{ for } t \geq \tau_0.$$

We can take Δ_0 so small that

$$N_{2\Delta_0}(\overline{K^{(i)}} \cup \phi(\overline{K^{(i)}})) \cap \Psi^{(j)} = \emptyset.$$

Find $k_1 > 0$ such that for $k \geq k_1$ we have

$$d(\phi^{\alpha_k}(\xi_k), \Psi^{(j)}) < \Delta_0.$$

There is a function $\tau_1(k)$ such that

$$\lim_{k\to\infty} \tau_1(k) = \infty, d(\phi^{\tau_k^{(1)}+1+t}(\xi_k), \phi^t(z_1)) < \Delta_0$$

for $0 \leq t \leq \tau_1(k)$, hence $\tau_k^{(1)} + 1 + \tau_k < \alpha_k$. Evidently we can take $\tau_1(k)$ such that we have

$$\phi^{\tau_k^{(1)}+1+t}(\xi_k) \in K^{(i)} \text{ if } 0 \leq t \leq \tau_1(k),$$

so we obtain a contradiction with the definition of $\tau_k^{(1)}$. This proves (.9) .

As $t \to +\infty$ the point $\phi^t(z_1)$ tends to one of the sets Ψ_b , $1 \leq b \leq \mu$, $\Psi_b \neq \Psi^{(i)}$. Denote this set $\Psi^{(i_2)}$. If $i_2 = j$ this completes the construction of a collection $z_1, \ldots, z_\sigma$. Otherwise it follows from the construction of z_1 that there exists a number Δ_1 and sequences $\xi_k^{(2)} \in O^+(\xi_k)$, $\alpha_k^{(2)}$ such that

$$\lim_{k\to\infty} d(\xi_k^{(2)}, \Psi^{(i_2)}) = 0, \lim_{k\to\infty} d(\phi^{\alpha_k^{(2)}}(\xi_k^{(2)}), \Psi^{(j)}) = 0,$$

$$\lim_{k\to\infty} \alpha_k^{(2)} = +\infty, d(\phi^t(\xi^{(2)}), \Psi^{(i)}) \geq \Delta_1 > 0$$

for $t \in [0, \alpha_k^{(2)}]$. Repeating the arguments we used to construct z_1 we can find a point

$$z_2 \in (X \setminus W^s(\Psi^{(i)})) \cap (W^u(\Psi^{(i_2)}) \setminus W^s(\Psi^{(i_2)})).$$

Find $i_3 \neq i, i_2$ such that

$$z_2 \in W^s(\Psi^{(i_3)})$$

and so on. As the collection $\Psi_1, \ldots, \Psi_\mu$ is finite the described process produces points $z_1, \ldots, z_\sigma$ with desired properties.

Lemma B.12. *Let x^* be a point belonging to $D(I) \cap M_0$. There exists a set Ψ_i such that Ψ_i has the property B and*

$$x^* \in W^s(\Psi_i).$$

Proof. Let X be the ω-limit set of $O(x^*)$. Evidently X is a compact subset of J . Assume that $\Psi^{(1)}, \ldots, \Psi^{(m)}$ are all the sets in the collection $\Psi_1, \ldots, \Psi_\mu$ which have non-empty intersections with X . Find $\Delta_0 > 0$ such that for any $x \in \Psi$ with $x \notin \Psi^{(1)} \cup \ldots \cup \Psi^{(m)}$ we have $d(x, X) > \Delta_0$.

Suppose that $m > 1$, that is that X does not belong to a single set Ψ_i . Take two distinct sets $\Psi^{(i)}, \Psi^{(j)}$, $1 \leq i, j \leq m$. Fix a sequence of points $\xi_k \in O(x^*)$ with

$$\lim_{k\to\infty} d(\xi_k, \Psi^{(i)}) = 0,$$

and a sequence of numbers α_k with

$$\lim_{k\to\infty} \alpha_k = \infty, \lim_{k\to\infty} d(\phi^{\alpha_k}(\xi_k), \Psi^{(j)}) = 0.$$

As

$$\lim O_0^{\alpha_k}(\xi_k) \subset X \subset J$$

we can apply Lemma B.11 to construct points $z_1, \ldots, z_{\sigma-1}$ and sets $\Psi^{(i_s)}$.

Note that if p, q are points of hyperbolic trajectories and $p \to q$ then $\dim W^u(p) \geq \dim W^u(q)$ (see [Pi8], Chap.6), hence the function $u(x) = \dim W^u(x)$ equals to a constant on Ψ_i . Let u_i be this constant.

Take a point $z_a, a \in \{1, \ldots, \sigma\}$, and apply Lemma B.7 to find points p_a, q_a such that

$$z_a \in W^u(p_a) \cap W^s(q_a), O(p_a) \subset K^{(i_a)}, O(q_a) \subset K^{(i_{a+1})}.$$

As $\dim W^u(p_a) = u_{i_a}$, and z_a is a point of transversal intersection of $W^u(p_a)$, $W^s(q_a)$ (we take into account here that $z_a \in J$ and that ϕ satisfies the STC on J) we see that $u_{i_a} \geq u_{i_{a+1}}$. Hence, $u_i \geq u_j$. The indices $i, j \in \{1, \dots, m\}$ are arbitrary, we obtain that $u_1 = \dots = u_m$.

For any point $x \in X$ we can find points $x_- \in K^{(i)}, x_+ \in K^{(j)}$; $i, j \in \{1, \cdots, m\}$ such that x is a point of transversal intersection of $W^u(x_-), W^s(x_+)$. It follows from Theorem 2.1 of Chap.4 in [Pl1] that the trajectory $O(x)$ is hyperbolic, hence the set X is hyperbolic, and $X \subset H(C_1, \lambda_1)$ (see Lemma B.1) . The same arguments as in the proof of Lemma B.7 show that there is a point γ with $O(\gamma) \subset H(C, \lambda)$ and such that

$$x^* \in W^s(\gamma), r(\overline{O(\gamma)}, X) < \Delta_1 = \min(\Delta, \Delta_0).$$

Let X^+ be the ω-limit set of $O(\gamma)$, fix a point $x^+ \in X^+$. As the set X^+ is hyperbolic there is a sequence $\pi_m \in \Pi$ such that

$$x^+ \in \lim \pi_m, r(\pi_m, X) < \Delta_1.$$

Take a small disc d_+ in $W^u(x_+)$ containing x^+. There exists a sequence of discs $d_k \subset W^u(O(\gamma))$ with

$$\lim_{k \to \infty} \rho_1(d_+, d_k) = 0.$$

Choose m, k_0 such that $W^s(\pi_m)$ has points of transversal intersection with d_+ and with $d_k, k \geq k_0$. As M_0 is open, M_0 is transversal to $W^s(\gamma)$ at x^* , hence there exists a sequence of discs $a_k \subset M_0$ such that

$$\lim_{k \to \infty} \rho_1(a_k, d_k) = 0.$$

It follows that we can find k with

$$a_k \cap W^s(\pi_m) \neq \emptyset.$$

Apply the λ-Lemma to show that $\pi_m \in J$. By the choice of Δ_0 there is a set $\Psi^{(i)}, i \in \{1, \dots, m\}$ such that $\pi_m \in \Psi^{(i)}$, that is $\Psi^{(i)}$ has the property B. The existence of points $z_1, \dots, z_\sigma$, and Lemma B.10 imply that any set $\Psi^{(j)}, j \in \{1, \dots, m\}$ has the property B. Now it follows from Lemma B.9 that $m = 1$, that is there is a single set Ψ_i such that $X \cap \Psi_i \neq \emptyset$.

Let us show that $X \subset \Psi_i$. Fix a point $x \in X$, and take points $x^+, x^- \in \Psi_i$ such that

$$x \in W^u(x^-) \cap W^s(x^+).$$

We can construct a transversal homoclinic contour $\Gamma \subset J$ and hyperbolic periodic points π_m with

$$x \in \lim \pi_m, \pi_m \cap K_i \neq \emptyset$$

(see Lemma B.9). For large m we have $\pi_m \in \Psi_i$, hence $x \in \Psi_i$. This completes the proof.

Lemma B.13. *Every set in the collection $\Psi_1, \ldots, \Psi_\mu$ has the property B.*

Proof. Fix a set Ψ_i , $i \in \{1, \ldots, \mu\}$, and consider a sequence $z_k \in M_0$ such that

$$\lim_{k\to\infty} d(z_k, \Psi_i) = 0.$$

We may consider points $z_k \in D(I)$. Find for any point z_k a number t_k such that

$$z_k = \phi^{t_k}(x_k), x_k \in V_0$$

(the compact set V_0 was defined after Lemma B.1). It is evident that $\lim_{k\to\infty} t_k = +\infty$.

Let p_0 be a limit point of the sequence x_k . Apply Lemma B.12 to find a set $\Psi^{(1)}$ such that $\Psi^{(1)}$ has the property B and $p_0 \in W^s(\Psi^{(1)})$. If $\Psi^{(1)} = \Psi_i$ there is nothing to prove.

If $\Psi^{(1)} \neq \Psi_i$ we can find $\Delta_0 > 0$ such that

$$d(\phi^t(p_0), \Psi_i) \geq \Delta_0 \text{ for } t \geq 0.$$

It is easy to see that there exist numbers $\tau_k > 0$ such that

$$d(\phi^t(p_0), \phi^t(x_k)) < \frac{\Delta_0}{2} \text{ for } 0 \leq t \leq \tau_k,$$

and if $\xi_k = \phi^{\tau_k}(x_k)$ then

$$\lim_{k\to\infty} d(\xi_k, \Psi^{(1)}) = 0.$$

In this case $\alpha_k = t_k - \tau_k > 0$ for large k , and we can apply Lemma B.11 (taking $\Psi^{(1)}, \Psi_i$ as $\Psi^{(i)}, \Psi^{(j)}$) to find sets $\Psi^{(2)}, \ldots, \Psi^{(\sigma)} = \Psi_i$ such that

$$W^u(\Psi^{(j)}) \cap W^u(\Psi^{(j+1)}) \neq \emptyset, j = 1, \ldots, \sigma - 1.$$

As $\Psi^{(1)}$ has the property B it follows from Lemma B.10 that Ψ_i has the property B.

Lemma B.14. $\Omega(J) = \Psi$.

Proof. It is evident that $\Psi \subset \Omega(J)$. Let us show that

$$\Omega(J) \subset \Psi. \tag{.10}$$

Take a point $x_0 \in \Omega(J)$. It follows from the definition of the set Ψ and from Lemmas B.8,B.13 that we can find sets $\Psi^{(0)}, \Psi^{(1)}$ (in the collection $\Psi_1, \ldots, \Psi_\mu$) , and points $x \in \Psi^{(0)}$, $y \in \Psi^{(1)}$ such that

$$x_0 \in W^u(x) \cap W^s(y).$$

Suppose that $\Psi^{(0)} = \Psi^{(1)}$. Find sequences $p_k, q_k \in \Pi$ such that $p_k \Rightarrow x, q_k \Rightarrow y$. It was shown in the proof of Lemma B.8 that in this case for large k we have $p_k, q_k \in \Psi^0$ and there exist points y_k of transversal intersection of $W^s(p_k), W^u(q_k)$, and points z_k of transversal intersection of $W^u(p_k), W^s(q_k)$ such that $\lim_{k\to\infty} z_k = x_0$. Hence for large k we obtain transversal homoclinic contours

$$\Gamma_k = \overline{O(y_k)} \cup \overline{O(z_k)}, \Gamma_k \subset J.$$

It follows that these $\Gamma_k \subset H(C_1, \lambda_1)$. By Lemma B.5 there exist points $r_k \in \Pi$ with

$$x_0 \in \lim r_k, \lim_{k\to\infty} r(O(r_k), J) = 0.$$

Then $r_k \in \Psi^{(0)}$ for large k , so that $x_0 \in \Psi^{(0)} \subset \Psi$.

If $\Psi^{(0)} \neq \Psi^{(1)}$ let us fix sequences of points z_k and of numbers t_k such that

$$\lim_{k\to\infty} z_k = x_0, \lim_{k\to\infty} \phi^{t_k}(z_0) = x_0, \lim_{k\to\infty} t_k = \infty.$$

Evidently there exist sequences τ_k, θ_k having the following property : $0 < \tau_k < \theta_k < t_k$, and if $\alpha_k = \theta_k - \tau_k$, $\xi_k = \phi^{\tau_k}(z_k)$ then

$$\lim_{k\to\infty} d(\xi_k, \Psi^{(1)}) = 0, \lim_{k\to\infty} \alpha_k = \infty, \lim_{k\to\infty} d(\phi^{\alpha_k}(\xi_k), \Psi^{(0)}) = 0.$$

Now we can apply the same proof as in Lemma B.11 to show that there exist sets $\Psi^{(2)}, \ldots, \Psi^{(\sigma)} = \Psi^0$ such that

$$W^u(\Psi^{(j)}) \cap W^s(\Psi^{(j+1)}) \neq \emptyset, j = 1, \ldots, \sigma - 1.$$

Note that the condition

$$\lim_{k\to\infty} r(O_0^{\alpha_k}(\xi_k), J) = 0$$

was used to show that $z_1 \in J$ (after (.9)). In this case the same reasons as in the proof of (.9) show that $z_1 \in W^u(\Psi^{(1)})$. Now we know that $\Psi^{(1)}$ has the property B, it follows that $W^u(\Psi^{(1)}) \subset J$, hence $z_1 \in J$.

After that we can construct a sequence of transversal homoclinic contours $\Gamma_k \subset J$ and a sequence of hyperbolic periodic points $r_k \in J$ with $\lim_{k\to\infty} r_k = x_0$ as in the first case. Again we obtain $x_0 \in \Psi$.

This proves (.10) and completes the proof of our lemma.

It follows now from Lemmas B.1,B.14 that the set $\Omega(J)$ is hyperbolic. It was noticed earlier that for any point $p \in \Psi$ there is a sequence of periodic points x_k such that $\lim_{k\to\infty} x_k = p$. By Lemmas B.8,B.13 in this case for large k we have $x_k \in \Psi$. Now Lemma B.14 implies that periodic points of ϕ which belong to J are dense in $\Omega(J)$. This proves the statement (b) of Theorem B.1.

The statement (c) is a consequence of (b) and of the STC on J.

For any point $x \in J$ there is a point $p \in \Psi = \Omega(J)$ with $x \in W^u(p)$ (see Lemmas B.8,B.13). This establishes (d).

To decompose $\Omega(J)$ in the form (.1) take $\Omega_i = \Psi_i$. The reasons used in the proof of the Spectral Decomposition Theorem [Sm2] show that as the sets Ω_i are hyperbolic and periodic points are dense in Ω_i , any of these sets contains a dense trajectory.

So it remains to prove that J is an attractor. Of course, this statement is not trivial only in the case $\text{Int}I \neq \emptyset$, so we consider below this case.

First let us show that the set J is Lyapunov stable. Fix a sequence ϵ_j of positive numbers with $\lim_{j\to\infty}\epsilon_j = 0$. Denote

$$D_j = \{x \in \text{Int}I : d(x, J) \leq \epsilon_j\},$$

$$E_j = \overline{(\text{Int}I) \setminus D_j}.$$

To obtain a contradiction suppose that J is not Lyapunov stable. Then we can find j_0 such that for $j \geq j_0$ the following is true. There exists a sequence of points z_i (depending on j) such that

$$\lim_{i\to\infty} d(z_i, J) = 0, O^+(z_i) \cap E_j \neq \emptyset.$$

Find for any z_i a number $\tau_i > 0$ such that

$$d(\phi^t(z_i), J) < \epsilon_j \text{ for } 0 \leq t < \tau_i;$$

$$\xi_i = \phi^{\tau_i}(z_i) \in E_j.$$

Let q_j be a limit point of the sequence ξ_i . It follows from the construction that $\phi^t(q_j) \in D_j$ for $t < 0$. Denote by X_j the α-limit set of $O(q_j)$. Evidently, $X_j \subset \Omega(\phi)$ and $X_j \subset D_j$. Let

$$X = \lim_{j\to\infty} X_j,$$

then $X \subset J$ and $X \subset \Omega(J)$. Hence, we can find j_1 such that for $j \geq j_1$ we have $X_j \subset W$, and $X_j \subset H(C, \lambda)$.

We claim that if j is large enough then

$$X_j \subset J. \tag{.11}$$

Suppose that there are arbitrarily large j such that $X_j \setminus J \neq \emptyset$. As the sets X_j are hyperbolic for $j \geq j_1$ we can find periodic points $p_j \in \Pi \setminus J$ with

$$r(O(p_j), J) < 2\epsilon_j.$$

Let p be a limit point of this sequence p_j . Then $p \in \Omega(J) = \Psi$, hence $p \in \Psi_i$ for some $i \in \{1, \ldots, \mu\}$. It was shown earlier (see Lemmas B.8,B.13) that in this case $p_j \in \Psi_i$ for large j. Therefore $X_j \subset \Psi$ for these j. The same reasons as in the proof of Lemma B.12 show that there is a set Ψ_i with $X_j \subset \Psi_i$. This implies

$$q_j \in W^u(\Psi_i),$$

so we obtain a contradiction between $W^u(\Psi_i) \subset J$ (see Lemmas B.8,B.13) and $q_j \notin J$. This contradiction shows that J is Lyapunov stable.

Let us show that there exists a neighborhood U of J such that for $x \in U$ we have

$$\lim_{k\to\infty} d(\phi^k(x), J) = 0.$$

Consider again a sequence $\epsilon_j > 0$, $\lim_{j\to\infty} \epsilon_j = 0$. If no neighborhood of J has the described property then we can find points x_j such that

$$r(O^+(x_j), J) \leq \epsilon_j$$

(we take into account here that J is Lyapunov stable), and $d(\phi^k(x_j), J)$ does not tend to zero as $k \to +\infty$.

Let X_j be the ω-limit point of $O(x_j)$. These sets have the following properties : X_j are invariant and compact, $X_j \subset \Omega(\phi)$, $X_j \setminus J \neq \emptyset$, and $r(X_j, J) \leq \epsilon_j$. By the previous argument, the existence of a sequence of sets X_j with these properties leads to a contradiction. This completes the proof of Theorem B.1.

Appendix C. Complete Families of δ-
-semi-trajectories.

Fix a system $\phi \in Z(M)$ and $\delta > 0$. We say that

$$\Xi = \{\xi(p) : p \in M\}$$

is a *complete family of* δ*-semi-trajectories* for ϕ ($CF(\delta, \phi)$ below) if for any $p \in M$

$$\xi(p) = \{p_k : k \geq 0\}$$

is a δ-semi-trajectory for ϕ with $p_0 = p$.

The main reason to investigate $CF(\delta, \phi)$ is the following one. We may consider a numerical method of accuracy $\delta > 0$ for our system ϕ as a map $\psi : M \to M$ such that

$$d(\phi(x), \psi(x)) < \delta \tag{.1}$$

for any $x \in M$.

Assume that (.1) is satisfied. Take a point $p \in M$ and construct the "numerical trajectory"

$$p_0 = p, p_1 = \psi(p), \dots, p_k = \psi^k(p), \dots.$$

Of course, $\xi(p) = \{p_k : k \geq 0\}$ is a δ-semi-trajectory for ϕ. Constructing "numerical trajectories" of this type through all points of M we obtain a $CF(\delta, \phi)$.

Note that it is natural to consider numerical methods ψ being neither continuous nor invertible.

An important problem is the problem of correspondence between real and numerical trajectories. If our system ϕ has the POTP$^+$ (see section 1.2) then given $\epsilon > 0$ we can find $\delta > 0$ such that for any approximate semi-trajectory $\xi(p) = \{\psi^k(p) : k \geq 0\}$ obtained by a numerical method ψ of accuracy δ there is a real trajectory $\{x_k = \phi^k(x)\}$ with

$$d(x_k, p_k) < \epsilon, k \geq 0.$$

This means that if a system ϕ has the POTP$^+$ then the qualitative picture of trajectories obtained by a numerical method ψ of good accuracy is properly reflected by a part of the qualitative picture of real trajectories. Theorem 1.2.1$'$ shows that a generic system in CLD(M) (and a generic system in $Z(M)$ if dim$M \leq 3$) has the described property. It follows from Theorem A.1 that if ϕ is a diffeomorphism which satisfies the STC then the same is true (with $\epsilon = L\delta$ for some $L > 0$).

It was shown in Theorem 1.2.2′ that a generic system ϕ in $Z(M)$ (with arbitrary dim$M$) has the following property : given $\epsilon > 0$ we can find $\delta > 0$ such that for any approximate semi-trajectory $\xi(p) = \{\psi^k(p) : k \geq 0\}$ obtained by a numerical method of accuracy δ there is a point x with

$$\xi(p) \subset N_\epsilon(O^+(x, \phi)).$$

In this case we also can be sure that the qualitative picture obtained by a numerical method of good accuracy gives information about the real qualitative picture.

Now let us discuss the following question : which information about the complete set of trajectories of a given system ϕ can be obtained by numerical methods of good accuracy? This question leads us to investigate properties being "inverse" to ones discussed above. Here we describe some results obtained by R.Corless and the author in [Cor].

Let us show that for any system $\phi \in Z(M)$ the following statement being "inverse" to the statement of Theorem 1.2.2′ is true.

Theorem C.1. *Let ϕ be a dynamical system. Given $\epsilon > 0$ there exists $\delta > 0$ such that if $\Xi = \{\xi(p) : p \in M\}$ is a $CF(\delta, \phi)$ then for every $O^+(\phi)$ we can find $\xi(p) \in \Xi$ with*

$$O^+(\phi) \subset N_\epsilon(\xi(p)).$$

Proof. Fix a dynamical system $\phi \in Z(M)$ and $\epsilon > 0$. Consider a finite open covering $\{M_i\}$, $i = 1, \dots, k$ of M such that diam$M_i < \epsilon$, $i = 1, \dots, k$. Let $K = \{1, \dots, k\}$. Consider the set L of subsets $\lambda \subset K : \lambda = \{l_1, \dots, l_m\} \subset K$ is in L if and only if there exists a semi-trajectory $O^+(\phi)$ such that

$$O^+(\phi) \subset \bigcup_{i \in \lambda} M_i; O^+(\phi) \cap M_i \neq \emptyset, i \in \lambda. \tag{.2}$$

Take $\lambda \in L$ and $p \in M$ such that (.2) is satisfied for $O^+(\phi) = O^+(p, \phi)$. Let $\lambda = \{l_1, \dots, l_m\}$. There exist $k_1, \dots, k_m$ such that

$$\phi^{k_i}(p) \in M_{l_i}, i = 1, \dots, m.$$

Evidently there is $\delta = \delta(\lambda)$ having the following property : if $\xi(p) = \{p_k : k \geq 0\}$ is a δ-semi- trajectory with $\delta \in (0, \delta(\lambda))$, and $p_0 = p$ then

$$p_{k_i} \in M_{l_i}, i = 1, \dots, m.$$

Hence,

$$O^+(p, \phi) \subset N_\epsilon(\xi(p)).$$

The set L is finite, take

$$\delta = \min_{\lambda \in L} \delta(\lambda).$$

Evidently this δ has the required property. This completes the proof.

Now let us consider the following property of "tracing of real trajectories by approximate ones" being "inverse" to the POTP$^+$ for $\phi \in Z(M)$: given $\epsilon > 0$ there exists $\delta > 0$ such that for any $\Xi = \{\xi(p) : p \in M\}$ which is a $CF(\delta, \phi)$, and for any $x \in M$ there is $\xi(p) = \{p_k : k \geq 0\}$ with

$$d(\phi^k(x), p_k) < \epsilon, k \geq 0.$$

We show that diffeomorphisms which satisfy the STC do not have this property. To be exact, we establish the following result.

Theorem C.2 *Let ϕ be a diffeomorphism of class C^1 which satisfies the STC. We can find $a > 0$ and a periodic point q of ϕ which have the following property. For any $\delta > 0$ there is $\Xi = \{\xi(p) : p \in M\}$ being a $CF(\delta, \phi)$ and such that for any $\xi(p) \in \Xi$ we have*

$$\xi(p) \not\subset N_a(O(q, \phi)).$$

Proof. Let us begin with the case of a Morse-Smale diffeomorphism ϕ. In this case the nonwandering set $\Omega(\phi)$ consists of a finite number of periodic points. There exists a periodic point q such that $Q = O(q, \phi)$ is a source. That means that the stable manifold $W^s(Q) = Q$, and the unstable manifold $W^u(Q)$ contains a neighborhood of Q (in other words Q is an attractor for ϕ^{-1}). Let

$$a = \frac{1}{2} \min_{r \in Q, p \in \Omega(\phi) \setminus Q} d(r, p),$$

and let L be a Lipschitz constant for ϕ. Fix arbitrary $\delta > 0$. Find a point $y \in W^u(Q)$ such that

$$y \notin Q, d(q, y) < \frac{\delta}{L}.$$

Define ψ as follows : $\psi(x) = \phi(x)$ for $x \neq q$, $\psi(q) = y$. Then evidently (.1) holds. It is easy to see that for any $x \in M$ we have

$$\psi^k(x) \to \Omega(\phi) \setminus Q$$

as $k \to \infty$. So if we take $\Xi = \{\xi(p) : p \in M\}$ where $\xi(p) = \{\psi^k(p) : k \geq 0\}$ then Ξ is a $CF(\delta, \phi)$, and for any $\xi(p) \in \Xi$ we have

$$\xi(p) \not\subset N_a(Q).$$

Now consider the case when the set $\Omega(\phi)$ is infinite. Let

$$\Omega(\phi) = \Omega_1 \cup \cdots \cup \Omega_m$$

be the decomposition into basic sets. Take an infinite basic set Ω_j . It is well-known that for any trajectory Q in Ω_j its stable manifold $W^s(Q)$ is dense in Ω_j .

As periodic points are dense in Ω_j we can find two distinct periodic trajectories Q_1, Q_2 in Ω_j . Evidently there exists $a > 0$ such that for any point $q \in Q_1$ we have

$$d(p,q) \geq 2a, p \in Q_2;$$
$$d(q,x) \geq 2a, x \in \Omega_i, i \neq j.$$

Fix arbitrary $\delta > 0$ (we consider $\delta < a$) and let L be a Lipschitz constant of ϕ. Consider

$$V = N_{\frac{\delta}{2L}}(\Omega_j).$$

Define ψ as follows :
-for $x \in (M \setminus V) \cup W^s(Q_2)$ let $\psi(x) = \phi(x)$;
-for $x \in V \setminus W^s(Q_2)$ find $y \in W^s(Q_2)$ such that

$$d(x,y) < \frac{\delta}{L}$$

(this is possible as $W^s(Q_2)$ is dense in Ω_j), and let $\psi(x) = \phi(y)$.

Evidently, (.1) is satisfied. Take $\Xi = \{\xi(p) : p \in M\}$ where $\xi(p) = \{\psi^k(p) : k \geq 0\}$, then Ξ is a $CF(\delta, \phi)$.

Consider arbitrary $x \in M$. As ϕ satisfies the STC there is a unique basic set Ω_i such that

$$\lim_{k\to\infty} d(\phi^k(x), \Omega_i) = 0$$

(see Theorem 0.4.3). If $\phi^k(x) \notin V$ for $k \geq 0$ then

$$\xi(x) = \{\psi^k(x) : k \geq 0\} = \{\phi^k(x) : k \geq 0\},$$

and $\xi(x)$ contains points which are arbitrarily close to $\Omega_i \neq \Omega_j$, hence

$$\xi(x) \not\subset N_a(Q_1).$$

If $O^+(x,\phi) \cap V \neq \emptyset$, take $l \geq 0$ such that $\phi^k(x) \notin V$ for $k \leq l-1$; $z = \phi^l(x) \in V$. Then $\psi^k(z) \in W^s(Q_2)$ for $k \geq 0$, so that

$$\lim_{k\to\infty} d(\psi^k(x), Q_2) = 0,$$

and

$$\xi(x) \not\subset N_a(Q_1).$$

This completes the proof.

It follows from this theorem that for any diffeomorphism ϕ satisfying the STC there exist numerical methods of arbitrary accuracy such that ϕ has trajectories which are not weakly traced by approximate trajectories obtained using these methods.

References

[An1] Anosov, D.V.: Geodesic flows on closed Riemannian manifolds of negative curvature. Proc. Steklov Math. Inst. **90** (1967), AMS (1969)

[An2] Anosov, D.V.: Ob odnom klasse invariantnyh mnojestv gladkih dinamicheskih sistem. On a class of invariant sets of smooth dynamical systems (in Russian). In: Proc. 5th Int. Conf. on Nonl. Oscill. **2** Kiev (1970) 39–45

[Au] Auslander, J. and Seibert, P.: Prolongation stability in dynamical systems. Ann. Inst. Fourier, Grenoble **14** (1964) 237–268

[Bh] Bhatia, N.P. and Szegö, G.P.: Stability theory of dynamical systems. Springer-Verlag (1970)

[Boh] Bohl, P.:Über Differentialungleichungen. J. für reine und angew. Math. **144** (1913) 284–318

[Bow] Bowen, R.: Equilibrium states and the ergodic theory of Anosov diffeomorphisms. Lect. Notes in Math. **470** Springer-Verlag (1975)

[Con] Conley, R.: Isolated invariant sets and the Morse index. Reg. Conf. Series in Math., AMS, Providence, R.I. **38** (1978)

[Cor] Corless, R. and Pilyugin, S.Yu.: Approximate and real trajectories for generic dynamical systems (to appear)

[Cov] Coven, E.M., Madden, J. and Nitecki, Z.: A note on generic properties of continuous maps. In: Ergodic Theory and Dynamical Systems II (Progress in Math. **21**) Birkhäuser-Verlag (1982) 97–101

[D1] Dobrynsky, V.A. and Sharkovsky, A.N.: Tipichnost' dinamicheskih sistem dlya kotoryh pochti vse traektorii ustoichivy pri postoyanno deistvuyuschih vozmuscheniyah. Genericity of dynamical systems for which almost all trajectries are stable with respect to permanent perturbations (in Russian). Dokl.Ac.Nauk SSSR **211** (1973) 273–276

[D2] Dobrynsky, V.A.: Tipichnost' dinamicheskih sistem s ustoichivoy prolongatsiey. Genericity of dynamical systems with stable prolongation (in Russian). In: Dynamical Systems and Problems of Stability. Kiev (1973) 43–53

[Hi1] Hirsch, M.: Differential topology. Springer-Verlag (1976)

[Hi2] Hirsch, M., Palis, J., Pugh, C., and Shub, M.: Neighborhoods of hyperbolic sets. Invent. math. **9** (1970) 133–163

[Hu1] Hurley, M.: Attractors : persistence, and density of their basins. Trans. of the AMS **269** (1982) 247–271

[Hu2] Hurley, M.: Consequences of topological stability. J. Diff. Equat. **54** (1984) 60–72

[I] Ivanov, O.A. and Pilyugin, S.Yu.: Lipshitzeva R-ustoichivost' prityagivayuschih mnojestv. Lipschitz R-stability of attractive sets (in Russian). Diff. Uravn. **24** (1988) 776–784

[K] Kuratovski, K.: Topology. Ac.Press (1966)

[L] Lewowicz, J.: Lyapunov functions and topological stability. J. Diff. Equat. **38** (1980) 192–209

[Ma1] Mañé, R.: Characterizations of AS diffeomorphisms. In: Geom. and Top.III Lat. Am. Sch. of Math., July 1976, Lect. Notes in Math. **597** Springer-Verlag (1977) 389–394

[Ma2] Mañé, R.: A proof of the C^1-stability conjecture. IHES Publ. Math. **66** (1988) 161-210

[Mi] Milnor, J.: On the concept of attractor. Commun. Math. Phys. **99** (1985) 177–196

[More] Moreva, M.B.: Ustoichivost' prityagivayuschih mnojestv v metrike R_0 otnositelno C^0-vozmuschenii. Stability of attractive sets in the metric R_0 with respect to C^0-perturbations (in Russian). Vestnik Leningr. Univ. **15** (1988) 36–39

[Mori1] Morimoto, A.: Stochastically stable diffeomorphisms and Takens conjecture. Surikais Kokyuruko **303** (1977) 8–24

[Mori2] Morimoto, A.: The method of pseudo-orbit tracing and stability of dynamical systems. Sem. Note **39** Tokyo Univ. (1979)

[Mu1] Munkres, J.: Obstructions to the smoothing of piecewise-differentiable homeomorphisms. Ann. Math. **72** (1960) 521–554

[Mu2] Munkres, J.: Elementary differential topology. Annals of Math. Stud. **54** Princeton Univ. Press (1966)

[Ne] Newhouse, S.: Diffeomorphisms with infinitely many sinks. Topology **12** (1974) 9–18

[Ni1] Nitecki, Z.: Differentiable dynamics. MIT Press (1971)

[Ni2] Nitecki, Z.: On semi-stability of diffeomorphisms. Invent. math. **14** (1971) 83–122

[Ni3] Nitecki Z. and Shub,M.: Filtrations, decompositions, and explosions. Amer. J. Math. **97** (1975) 1029–1047

[O] Odani, K.: Generic homeomorphisms have the pseudo-orbit tracing property. Proc. AMS **110** (1990) 281–284

[Pa1] Palis, J.: On Morse-Smale dynamical systems. Topology **8** (1969) 385–404

[Pa2] Palis, J., Pugh, C., Shub, M. and Sullivan,D.: Genericity theorems in topological dynamics. In: Dynamical Systems - Warwick 1974, Lect. Notes in Math. **468** Springer- Verlag (1975) 239–250

[Pa3] Palis, J. and Takens, F.: Topological equivalence of normally hyperbolic dynamical systems. Topology **16** (1977) 336–346

[Pa4] Palis, J. and de Melo, W.: Geometric theory of dynamical systems. An introduction. Springer-Verlag (1982).

[Pa5] Palis, J.: On the C^1 Ω-stability conjecture. IHES Publ. Math. **66** (1988) 211–215

[Pe] Peixoto, M.: Structural stability on two-dimensional manifolds. Topology **1** (1962) 101–120

[Pi1] Pilyugin, S.Yu.: Granitsa prityagivayuschego mnojestva periodicheskoi A-sistemy. Boundary of an attracting set of a periodic A-system (in Russian). Diff. Uravn. **17** (1981) 1621–1629

[Pi2] Pilyugin, S.Yu.: Prityagivayuschie mnojestva so strogim usloviem transversalnosti na granitse. Attracting sets with the strong transversality condition on the boundary (in Russian). Diff.Uravn. **22** (1986) 1532–1539

[Pi3] Pilyugin, S.Yu.: C^0-vozmuscheniya prityagivayuschih mnojetsv i ustoichivost' granitsy. C^0-perturbations of attractive sets and stability of the boundary (in Russian). Diff.Uravn. **22** (1986) 1712–1719

[Pi4] Pilyugin, S.Yu.: Predelnye mnojestva traektorii oblastei v dinamicheskih sistemah. Limit sets of trajectories of domains in dynamical systems (in Russian). Funct. Anal. i ego Pril. **23** (1989) 82–83

[Pi5] Pilyugin, S.Yu.: Tsepnye prolongatsii v tipichnyh dinamicheskih sistemah. Chain prolongations in generic dynamical systems (in Russian). Diff. Uravn. **26** (1990) 1334–1337

[Pi6] Pilyugin, S.Yu.: Prostranstvo dinamicheskih sistem s C^0-topologiei. The space of dynamical systems with the C^0-topology (in Russian). Diff.Uravn.**26** (1990) 1659–1670

[Pi7] Pilyugin, S.Yu.: Predelnye mnojestva oblastei v potokah. Limit sets of domains in flows (in Russian). Proc. Leningr. Math. Soc. **1** (1991) 211–228

[Pi8] Pilyugin, S.Yu.: Introduction to structurally stable systems of differential equations. Birkhäuser- Verlag. (1992)

[Pl1] Pliss, V.A.: Integralnye mnojestva periodicheskih sistem differentsialnyh uravnenii. Integral sets of periodic systems of differential equations (in Russian). Moscow. (1977)

[Pl2] Pliss, V.A.: Ravnomerno ogranichennye resheniya lineynyh sistem differentsialnyh uravnenii. Uniformly bounded solutions of linear systems of differential equations (in Russian). Diff. Uravn.**13** (1977) 883-891

[Pl3] Pliss, V.A.: Ustoichivost' proizvolnoi sistemy po otnosheniyu k malym v smysle C^1 vozmuscheniyam. Stability of an arbitrary system with respect to C^1-small perturbations (in Russian). Diff. Uravn.**16** (1980) 1981-1982

[Pl4] Pliss, V.A.: Svyaz' mejdu razlichnymi usloviyami strukturnoi ustoichivosti. Connection between different conditions of structural stability (in Russian). Diff. Uravn.**17** (1981) 828-835

[Pl5] Pliss, V.A.: Raspolojenie ustoichivyh i neustoichivyh mnogoobrazii giperbolicheskih sistem (in Russian). Disposition of stable and unstable manifolds of hyperbolic systems. Diff.Uravn.**20** (1984) 779-785

[Po1] Pogonysheva, V.N.: Ustoichivost' prityagivayuschego mnojestva otnositelno metriki R_1. Stability of an attractive set with respect to metric R_1 (in Russian). Vestnik Leningr. Univ. **1** (1990) 108-109

[Po2] Pogonysheva, V.N.: Ustoichivost' predelnogo mnojestva otnositelno hausdorfovoi metriki. Stability of a limit set with respect to the Hausdorff metric (in Russian). VINITI, Moscow (1990), dep.N4240-B90

[Po3] Pogonysheva, V.N.: Zavisimost' predelnogo mnojestva ot oblasti. Dependence of a limit set on the domain (in Russian). VINITI, Moscow (1990), dep.N4501-B90

[Pu] Pugh, C.: The closing lemma. Amer. J. Math.**89** (1967) 956-1009

[Robb] Robbin, J.: A structural stability theorem. Ann. Math.**94** (1971) 447-493

[Robi1] Robinson, C.: Structural stability for C^1-diffeomorphisms. J. Diff. Equat.**22** (1976) 28-73

[Robi2] Robinson, C.: Stability theorems and hyperbolicity in dynamical systems. Rocky Mount. J. of Math.**7** (1977) 425-437

[Sa] Sawada, K.: Extended f-orbits are approximated by orbits. Nagoya Math, J.**79** (1980) 33-45

[Sc] Scherbina, N.V.: Nepreryvnost' odnoparametricheskih semeistv mnojestv. Continuity of one-parameter families of sets (in Russian). Dokl. Ac. Nauk SSSR **234** (1977) 327-329

[Sha] Sharkovsky, A.N., Kolyada, S.F., Sivak A.G. and Fedorenko, V.V.: Dinamika odnomernyh otobrajenii. Dynamics of one-dimensional maps (in Russian). Kiev (1989)

[Shu1] Shub, M. and Smale, S.: Beyond hyperbolicity. Ann. Math. **96** (1972) 587-591

[Shu2] Shub, M.: Structurally stable diffeomorphisms are dense. Bull. of the AMS. **78** (1972) 817-818

[Shu3] Shub, M. and Sullivan, D.: Homology theory and dynamical systems. Topology **14** (1975) 109-132

[Sm1] Smale, S.: Morse inequalities for a dynamical system. Bull. of the AMS. **66** (1960) 43-49

[Sm2] Smale, S.: Differentiable dynamical systems. Bull. of the AMS. **73** (1967) 747-817

[Sm3] Smale, S.: The Ω-stability theorem. Glob.Analysis Symp. in Pure Math. **14** (1970) 289-297

[Sm4] Smale, S.: Stability and isotopy in discrete dynamical systems. Salvador Symp.on Dyn.Systems, Univ. of Bahia. (1971) 527-530

[Ta1] Takens, F.: On Zeeman's tolerance stability conjecture. Manifolds - Amsterdam, 1970, Lect. Notes in Math. **197** Springer-Verlag (1971) 209-219

[Ta2] Takens, F.: Tolerance stability. Dynamical Systems - Warwick 1974, Lect. Notes in Math. **468** Springer-Verlag (1975) 293-304

[Th] Thom, R.: Structural stability and morphogenesis. Benjamin, Mass. (1975)

[Wa1] Walters, P.: Anosov diffeomorphisms are topologically stable. Topology **9** (1970) 71-78

[Wa2] Walters, P.: On the pseudo orbit tracing property and its relationship to stability. The Structure of Attractors in Dynamical Systems, Lect. Notes in Math. **668** Springer-Verlag (1978) 231-244

[WWh] White, W.: On the tolerance stability conjecture. Salvador Symp. on Dyn. Systems, Univ. of Bahia (1971) 663-665

[Wh] Whitehead, J.: Manifolds and transverse fields in Euclidean space. Ann. Math. **73** (1961) 154-212

[Y1] Yano, K.; Topologically stable homeomorphisms of the circle. Nagoya Math. J. **79** (1980) 145-149

[Y2] Yano, K.: Generic homeomorphisms of S^1 have the pseudo-orbit tracing property. J. Fac. Sci. Univ. Tokyo, Sect.IA Math. **34** (1987) 51-55

[Z] Zubov, V.I.: Ustoichivost' dvijeniya (metody Lyapunova i ih primeneniya). Stability of motion (Lyapunov methods and their applications) (in Russian). Moscow (1984)

Index

Vol. 1478: M.-P. Malliavin (Ed.), Topics in Invariant Theory. Seminar 1989-1990. VI, 272 pages. 1991.

Vol. 1479: S. Bloch, I. Dolgachev, W. Fulton (Eds.), Algebraic Geometry. Proceedings, 1989. VII, 300 pages. 1991.

Vol. 1480: F. Dumortier, R. Roussarie, J. Sotomayor, H. Żołądek, Bifurcations of Planar Vector Fields: Nilpotent Singularities and Abelian Integrals. VIII, 226 pages. 1991.

Vol. 1481: D. Ferus, U. Pinkall, U. Simon, B. Wegner (Eds.), Global Differential Geometry and Global Analysis. Proceedings, 1991. VIII, 283 pages. 1991.

Vol. 1482: J. Chabrowski, The Dirichlet Problem with L^2-Boundary Data for Elliptic Linear Equations. VI, 173 pages. 1991.

Vol. 1483: E. Reithmeier, Periodic Solutions of Nonlinear Dynamical Systems. VI, 171 pages. 1991.

Vol. 1484: H. Delfs, Homology of Locally Semialgebraic Spaces. IX, 136 pages. 1991.

Vol. 1485: J. Azéma, P. A. Meyer, M. Yor (Eds.), Séminaire de Probabilités XXV. VIII, 440 pages. 1991.

Vol. 1486: L. Arnold, H. Crauel, J.-P. Eckmann (Eds.), Lyapunov Exponents. Proceedings, 1990. VIII, 365 pages. 1991.

Vol. 1487: E. Freitag, Singular Modular Forms and Theta Relations. VI, 172 pages. 1991.

Vol. 1488: A. Carboni, M. C. Pedicchio, G. Rosolini (Eds.), Category Theory. Proceedings, 1990. VII, 494 pages. 1991.

Vol. 1489: A. Mielke, Hamiltonian and Lagrangian Flows on Center Manifolds. X, 140 pages. 1991.

Vol. 1490: K. Metsch, Linear Spaces with Few Lines. XIII, 196 pages. 1991.

Vol. 1491: E. Lluis-Puebla, J.-L. Loday, H. Gillet, C. Soulé, V. Snaith, Higher Algebraic K-Theory: an overview. IX, 164 pages. 1992.

Vol. 1492: K. R. Wicks, Fractals and Hyperspaces. VIII, 168 pages. 1991.

Vol. 1493: E. Benoît (Ed.), Dynamic Bifurcations. Proceedings, Luminy 1990. VII, 219 pages. 1991.

Vol. 1494: M.-T. Cheng, X.-W. Zhou, D.-G. Deng (Eds.), Harmonic Analysis. Proceedings, 1988. IX, 226 pages. 1991.

Vol. 1495: J. M. Bony, G. Grubb, L. Hörmander, H. Komatsu, J. Sjöstrand, Microlocal Analysis and Applications. Montecatini Terme, 1989. Editors: L. Cattabriga, L. Rodino. VII, 349 pages. 1991.

Vol. 1496: C. Foias, B. Francis, J. W. Helton, H. Kwakernaak, J. B. Pearson, H_∞-Control Theory. Como, 1990. Editors: E. Mosca, L. Pandolfi. VII, 336 pages. 1991.

Vol. 1497: G. T. Herman, A. K. Louis, F. Natterer (Eds.), Mathematical Methods in Tomography. Proceedings 1990. X, 268 pages. 1991.

Vol. 1498: R. Lang, Spectral Theory of Random Schrödinger Operators. X, 125 pages. 1991.

Vol. 1499: K. Taira, Boundary Value Problems and Markov Processes. IX, 132 pages. 1991.

Vol. 1500: J.-P. Serre, Lie Algebras and Lie Groups. VII, 168 pages. 1992.

Vol. 1501: A. De Masi, E. Presutti, Mathematical Methods for Hydrodynamic Limits. IX, 196 pages. 1991.

Vol. 1502: C. Simpson, Asymptotic Behavior of Monodromy. V, 139 pages. 1991.

Vol. 1503: S. Shokranian, The Selberg-Arthur Trace Formula (Lectures by J. Arthur). VII, 97 pages. 1991.

Vol. 1504: J. Cheeger, M. Gromov, C. Okonek, P. Pansu, Geometric Topology: Recent Developments. Editors: P. de Bartolomeis, F. Tricerri. VII, 197 pages. 1991.

Vol. 1505: K. Kajitani, T. Nishitani, The Hyperbolic Cauchy Problem. VII, 168 pages. 1991.

Vol. 1506: A. Buium, Differential Algebraic Groups of Finite Dimension. XV, 145 pages. 1992.

Vol. 1507: K. Hulek, T. Peternell, M. Schneider, F.-O. Schreyer (Eds.), Complex Algebraic Varieties. Proceedings, 1990. VII, 179 pages. 1992.

Vol. 1508: M. Vuorinen (Ed.), Quasiconformal Space Mappings. A Collection of Surveys 1960-1990. IX, 148 pages. 1992.

Vol. 1509: J. Aguadé, M. Castellet, F. R. Cohen (Eds.), Algebraic Topology - Homotopy and Group Cohomology. Proceedings, 1990. X, 330 pages. 1992.

Vol. 1510: P. P. Kulish (Ed.), Quantum Groups. Proceedings, 1990. XII, 398 pages. 1992.

Vol. 1511: B. S. Yadav, D. Singh (Eds.), Functional Analysis and Operator Theory. Proceedings, 1990. VIII, 223 pages. 1992.

Vol. 1512: L. M. Adleman, M.-D. A. Huang, Primality Testing and Abelian Varieties Over Finite Fields. VII, 142 pages. 1992.

Vol. 1513: L. S. Block, W. A. Coppel, Dynamics in One Dimension. VIII, 249 pages. 1992.

Vol. 1514: U. Krengel, K. Richter, V. Warstat (Eds.), Ergodic Theory and Related Topics III, Proceedings, 1990. VIII, 236 pages. 1992.

Vol. 1515: E. Ballico, F. Catanese, C. Ciliberto (Eds.), Classification of Irregular Varieties. Proceedings, 1990. VII, 149 pages. 1992.

Vol. 1516: R. A. Lorentz, Multivariate Birkhoff Interpolation. IX, 192 pages. 1992.

Vol. 1517: K. Keimel, W. Roth, Ordered Cones and Approximation. VI, 134 pages. 1992.

Vol. 1518: H. Stichtenoth, M. A. Tsfasman (Eds.), Coding Theory and Algebraic Geometry. Proceedings, 1991. VIII, 223 pages. 1992.

Vol. 1519: M. W. Short, The Primitive Soluble Permutation Groups of Degree less than 256. IX, 145 pages. 1992.

Vol. 1520: Yu. G. Borisovich, Yu. E. Gliklikh (Eds.), Global Analysis – Studies and Applications V. VII, 284 pages. 1992.

Vol. 1521: S. Busenberg, B. Forte, H. K. Kuiken, Mathematical Modelling of Industrial Process. Bari, 1990. Editors: V. Capasso, A. Fasano. VII, 162 pages. 1992.

Vol. 1522: J.-M. Delort, F. B. I. Transformation. VII, 101 pages. 1992.

Vol. 1523: W. Xue, Rings with Morita Duality. X, 168 pages. 1992.

Vol. 1524: M. Coste, L. Mahé, M.-F. Roy (Eds.), Real Algebraic Geometry. Proceedings, 1991. VIII, 418 pages. 1992.